THE AGE OF GENOMES

Tales from the Front Lines of Genetic Medicine

Steven Monroe Lipkin
with Jon R. Luoma

BEACON PRESS
BOSTON

Beacon Press
Boston, Massachusetts
www.beacon.org

Beacon Press books
are published under the auspices of
the Unitarian Universalist Association of Congregations.

19 18 17 16 8 7 6 5 4 3 2 1

This book is printed on acid-free paper that meets the uncoated paper ANSI/NISO specifications for permanence as revised in 1992.

Text design by Wilsted & Taylor Publishing Services

The case histories described in this book are all based on real events. However, to protect patient confidentiality, all names, personal identifiers, personal details, and personally identifiable characteristics have been changed. In two instances, a composite of details and events from more than one case history were combined in order to protect patient confidentiality.

Library of Congress Cataloging-in-Publication Data

Names: Lipkin, Steven Monroe, author. | Luoma, Jon R., contributor.
Title: The age of genomes : tales from the front lines of genetic medicine / Steven Monroe Lipkin ; with Jon R. Luoma.
Description: Boston : Beacon Press, [2016] | Includes bibliographical references and index.
Identifiers: LCCN 2015036455| ISBN 978-0-8070-7457-2 (hardcover : alk. paper) | ISBN 978-0-8070-7458-9 (ebook)
Subjects: | MESH: Genetic Diseases, Inborn—Case Reports. | Genetic Engineering—Case Reports. | Genetic Testing—Case Reports.
Classification: LCC RB155 | NLM QZ 50 | DDC 616/.042—dc23
LC record available at http://lccn.loc.gov/2015036455

To Gina, Rachel, Sophie, Alyson, Ethan, and Zoe,
my wonderful family, for their great patience and help
during the many hours I spent working on this book.

CONTENTS

Introduction

In each of our beginnings was our personal genome, created when we were less than a single cell and with us every moment since. Our genome has helped define what we have, and haven't, done with our lives. This precise DNA barcode, because of its enormous length and complexity, distinguishes each of us not only from the other eight billion people alive today on the planet, but also from each and every other of the twenty-something billion human beings who have ever lived since the dawn of humanity two hundred thousand years ago. Indeed, distinguishing each of us from every other living organism, whether flora, fauna, virus, or archebacteria for the past four billion years. What could be more personal than that?

Even when it's hot outside, we don't walk around naked. People are typically squeamish about public nudity because they don't want every detail of their reproductive anatomy put out on display for all the world to see, and prudence prevails. But this particular modesty only runs skin deep. What is more profoundly revealing than being naked in a crowd? Today, certainly one answer to this question, mechanistically more than skin deep and down to the core of our very being, is the sequence of our very own genome.

However, even if we generally don't like getting naked in public, we will when there's a good reason to do so. One place people typically do this is their doctor's office. We bare ourselves despite that slightly awkward feeling we have waiting there, exposed on the usually cold examining table, to allow our doctor to see our most intimate parts because we know this can help keep us healthy and so it's worth the discomfort.

Conversely, hiding some of these parts and refusing a doctor's examination could cause a lot of pain, or worse, in the end. This is essentially the same reason why people decide to share this intimacy *in silico* by giving their personal genome information to their doctor. Such information provides a tool to help navigate us to a healthy one hundred and fifty years or so of life. This is not astrology: this is bona fide biological science. But what does it all add up to?

About twenty years ago, having all a person's DNA fully analyzed (commonly referred to as whole genome sequencing) cost about the same as an aircraft carrier. Now, it costs less than a Vespa and doesn't require a prescription. How incredible and astonishing is this era that we live in? Whole genome sequencing has become an established technology to identify previously undiagnosed diseases in patients and their families, to find missed drug targets in cancers, or even to embark on a journey of self-discovery about our roots and how we each became the person we are today. There are currently serious academic discussions promoting Generation XX/XY—sequencing the genomes of all newborn American babies from a drop of heel blood—beginning in the next decade or so. There are also thought leaders table thumping for ending the century-old practice of doctors diagnosing disease primarily by signs and symptoms and instead retooling medicine into a practice based on genetic and biochemical mechanisms of ailments. These are promising and profound changes.

It takes about twenty thousand or so genes to make a human being. Somewhat humbling, this is near the same number of genes as the mustard weed plant, and less than double the number of genes to make tiny fruit flies that we hardly notice. Yet, this is our personal instruction manual, and perhaps eventually the key to bringing us back from beyond the grave centuries after we have passed.

When geneticists and genetic counselors look at a person's genome, we look for specific changes, called mutations, that we think should be there. One can think of a strand of DNA as being something like a sentence, albeit laid out on the familiar twisting double helix. Individual "letters" in the words of that sentence comprise pairs of only four nitrogen-rich compounds: adenine, guanine, cytosine, and thymine, respectively represented by the letters A, G, C, and T. Mutations can be a misspelling by a single such letter (referred to as a "base") of DNA, such as "porcxpine" instead of "porcupine," or reshuffled letters ("pucropine"), and additions or losses can stretch for millions of bases. This code ultimately defines each of our personal destinies.

Think of all these genes as divisible into multiple distinct piles, somewhat like sorted dirty laundry. Into the first hamper of gene mutations go the relatively few select ones often referred to as "actionable." Like stained clothes in your dirty laundry that need to be sorted out to go to the dry cleaners, the hamper of actionable gene mutations has the most precious and sought after contents. Actionable genes are the ones medicine can create a plan to fix, gene mutations with both a categorical imperative to inform patients what to do and a package insert telling doctors how to do it. If you have ten doctors in a room and ask them what we should do next with actionable gene mutations, all the hands go up in unison, each proudly and enthusiastically displaying medical literature about the best therapy for treating the genetic illness.

If we find a mutation in an actionable gene, the doctor and patient stand on a road looking out on the horizon. In front of them, the road bifurcates like the letter Y. There is a distinct choice of what can be done next. For example, the breast/ovarian cancer risk gene BRCA1, which spawned terabytes of journal articles and numerous clinical trials with guidelines about what the patients can decide to do (or not do, if they so choose) next to reduce their risk of developing cancer with surveillance imaging tests, chemopreventative drugs, or *previvor* surgery. These are the mutations that my patients anxiously hold their breath for, the ones we doctors (and the electronic medical record intelligent apps now monitoring and alerting us) will be drawn to first. These mutations are usually the reason why people go see a geneticist in the first place.

If a mutation in an actionable gene is found, a great deal can be accomplished, not only for the patient in front of you, but also for their kin, like water seeping deep down into the roots of a family tree forever into the future. After you articulate pros and cons of potential actions and anticipate reactions from the patient on what comes next, you have to discuss the what, how, and when—as well as whether, since the patient fully controls his or her genetic information—to contact the brothers and sisters, aunts and uncles, and children. An individual's choice whether or not to disclose personal genetic finding is always his or her own.

Occasionally, actionable gene mutations seem to emerge out of nowhere. These are often referred to as *incidentalomas* or (more recently) secondary findings—serendipity's evil twin. You go in looking for the etiology of hereditary colorectal cancer to detect tumors early and save a life and you find . . . Gaucher disease? Where in the world did that come from? The revelation comes without a history, but with hope to improve the family's future: you have discovered the seeds of a treatable disorder for which there exists enzyme replacement therapy, a clear actionable plan, but for which the person sitting in front of you has none of its signs or symptoms—at least not yet? With this knowledge, you can prevent disease, pain, and suffering before it happens.

A second hamper holds the *wish-they-were-actionable* gene mutations, the unactionables. These terrible mutations portent a grim future, but medicine has, sadly, little to offer—knowledge without power, impotence. Many people just do not want to know what tragedy awaits them until it unfolds. An example is familial Alzheimer disease, for which we have little to offer patients who are still cogent and able to anticipate the future. Here, knowing what *may* be can cause great anxiety that is not productive—even, at worst, inspiring despair. Furthermore, most mutations in this group are for diseases for which people are at increased risk but not certain to develop, so that knowledgeable previvors end up anxiously looking decades hence. In such cases, it is very reasonable to argue, it may be better just not to know and reject what the genome era has to offer.

On the other hand, some people want to know in order to conduct their lives accordingly. Maybe take that around-the-world-cruise sooner rather than later? Maybe pick up the kids from school for the wonderful simple pleasures of ice cream instead of working late again to make partner? Fortunately, sometimes these genes are later tossed into the first hamper. For example, Gorlin syndrome is a genetic disorder that can cause hundreds of skin tumors to grow all over the body. Mutations cause skin cells to grow uncontrollably, like the sorcerer's apprentice. Today a new drug can treat this condition by inhibiting the signals emanating from the broken gene. A remarkably large army of smart, dedicated, underpaid and sleep-deprived medical researchers and medical professionals is constantly working to find new ways to toss genes from the second hamper into the first.

A third hamper contains the potentially actionable genes. Say you have a genetic variant that causes increased sensitivity to clopidogrel, a drug that you've never heard of, but will ten years hence. The reason you've never heard of it is that it is used to treat heart attacks, and you've never had a heart attack. But if you did suffer a cardiac even in the future, knowing about this genetic variant beforehand might save you from the agony of slowly bleeding to death. The potentially actionable genes are like something you might store in a box in the back of the garage—you may not use them often, but you probably wouldn't want to throw them out. Some patients will want to know about these, and others will not. Personal preference and general anxiety level play a role in whether this information is disclosed or not.

A fourth hamper has laundry thrown in that is really someone else's and not yours. These are genes that don't directly affect your health, but that may be more significant for your siblings, kids, and cousins. My own experience is that many people who come in for genetic testing are really doing it for their children's sake, not their own. They love their families more than they do themselves. They are everyday heroes.

We each carry more than one hundred potential mutations in one gene or another. Many are in genes that don't seem to be very important. Others don't have strong effects if you have one good copy and one bad copy, because disease will occur only if you have two bad copies. For example, you may have one bad copy of a mutation in a gene causing cystic fibrosis (CF), a debilitating disease of the lung and pancreas. In people of mixed European ancestry, about one out of every twenty-three people is running around the streets of New York City carrying one mutated copy of the CF gene without a clue that they have it. If you are done having children, knowing this doesn't affect you much and is all well and good. However, if your son or niece is planning a family, your genetic testing could be very useful information for them, because their partner could be tested specifically for CF mutations. If so, various approaches, including exciting technologies such as pre-implantation genetic diagnosis—and likely gene editing, coming soon—could be used to reduce your son or niece's risk of having a child with full-blown CF.

A fifth hamper holds genes that offer personal insights but aren't really interesting to anyone except you and your family, the laundry regulars. These are not necessarily mutations, but mostly more subtle changes called variants that can affect, often modestly, your weight, height, the age you began puberty, and the fabric of your daily life. Maybe they explain why you could never make that high C in chorus and sat in the back row, or why you almost failed high school math but had perfect pitch.

Think of the genes in this fifth hamper as "self-knowledge" genes. They don't go in the medical record, but knowing about them can change the way you look at yourself in the mirror every day of your life. These genes are often somewhat at odds with family lore. Many ethnic groups carry distinctive genetic signatures. For example, from a genome sequence you can usually tell if an individual is African American, Caucasian, Asian, Satnami, or Ashkenazi Jewish, even if you've never laid eyes on the patient. A well-regarded research scientist whom I had never met made his genome sequence publically available as part of a research study. I remember scrolling through his genetic variant files and trying, more successfully than I had expected, to guess what he would look like before I peeked at his webpage photo. The personal genome is more than skin deep.

I also recall a very awkward conversation in a case where I had counseled a family about a child with CF and inadvertently discovered the boy's father was . . . well, not his genetic father. This is called nonpaternity, and in fact arises in a few people out of every hundred who have single gene or whole genome testing.

A sixth hamper holds the mutations that are not yours and you want to avoid at all costs. They are like the bright red shirt that runs and ruins the rest of your laundry. These are the gene mutations that appear in tumors. Today we know that cancer is really a disease of DNA, and whole genome sequencing can find the very root of the cancer, with the hope of digging it out, at least sometimes.

Searching for mutations in developing cancers is now one of the most common reasons why people have their genome sequenced. These individuals are looking for actionable (treatable by drugs already on the market) or potentially actionable (treatable with drugs in clinical trials but not yet fully vetted) mutations in the tumor that are not in normal cells. The lists of genetic mutations in cancers, but not the rest of our bodies, are extremely long, running in the thousands typically. However, most of these are not actionable mutations.

The positive yield of genetic-testing searches for tumor actionable mutations is currently estimated to be in the high single digits, although there is great variation in these numbers among the different types of cancer, and the rate is getting higher every year as new drugs develop.

Along the way, somewhat like a two-for-one promotion, people who have their cancer genome fully sequenced have their own genome sequenced as well in parallel. This is done in order to tell which genetic variants are tumor specific, and which are inherited from your normal cells. Along this path are accidentally discovered secondary findings and examples from all hampers of genes.

Of course, this overview of the different types of information we get from whole genome sequencing is somewhat oversimplified. Twenty thousand genes can't be easily summarized in a few pages. If you printed a single human genome in twelve-point Times Roman font as one text line, it would stretch across the continental United States.

This book is for those adventurous, knowledge insatiable souls who want to dive into the depths of the human genome, millions of years of evolution, to get a better feel for the enormity and type of data that comes from having your genome sequenced. The chapters that follow share stories of brave people with genetic disease, set against the backdrop of contemporary New York City, one of the most diverse cities in the world in terms of genetic (ethnic) backgrounds and cultures, and one of the most economically divergent in the United States as well.

Because of its complex nature, there is tremendous confusion about genetic testing and genome sequencing. In today's atmosphere of great exuberance over the power of genetic knowledge and emerging revolutionary technologies, the combination of media hype, poor physician education on the subject of genetics, and a not-very-scientifically-literate American public also creates the potential for confusion and misinformation. The stories in this book aim to educate the reader about the great promise, pitfalls, and limitations of genetic testing in an era of affordable genome sequencing.

ONE

The Parisian Housekeeper

Sergeant Vincent was a former helicopter pilot with an active social life who came to see us at the San Diego Veteran's Administration Hospital Ambulatory Care Clinic. He was a Navy veteran and tough enough to survive Da Nang, but I thought he would faint (and I would have to call a code) when we told him the news.

"That girl gave me syphilis? Damn it, Doc, that is so nasty." Then, he cocked his head to the left, his weathered face glancing up at us with fright-dilated pupils. "So, what's going to happen to me? What do I do now?"

Giving good news to patients is a thousand times easier than bad news, and unfortunately ten times less common. I quickly reassured Sergeant Vincent that there were multiple drugs available to treat syphilis, that the pharmacy had them in stock, and that he could relax and anticipate that he would soon be cured—at least as long as he tried hard not to re-expose himself.

Syphilis, an infectious malady with a notorious history, was called the "French disease" in Italy, the "Italian disease" in France, the "Spanish disease" in the Netherlands, and so on. Although no nation seemed to want to claim its origins, the scourge was once as devastating as it was widespread.

During the Renaissance, an estimated million-plus Europeans were syphilitics, a substantial proportion of that continent's total population. The disease caused untold suffering and ignominy to men and women throughout the ages. Yet, today it occupies a relatively modest lot of real estate in the outer suburbs of medical textbooks. It can usually be cured easily with a single shot of antibiotics in the deltoid muscle of the arm.

For millennia, infectious diseases, particularly bacterial ones like the "French disease," were the most likely cause of death. At the dawn of the twentieth century, about one in three Americans died from infectious diseases, often as children. More than a century later, in the United States, the death toll from bacterial maladies has plummeted. We have effective, safe, and often cheap therapies to combat many of them, and this great progress is reflected in a dramatically reduced rate of death from infectious diseases, which now numbers less than 5 percent. Bacterial maladies like syphilis have evolved from hopeless tragedy to easily "actionable" diagnoses that would constitute medical malpractice if missed.

The Holy Grail of clinical genetics, which focuses on the diagnosis and medical management of maladies caused by genetic mechanisms, is to find similarly effective treatments for the roughly six to seven thousand genetic diseases that have arisen over the long sweep of human evolution. Collectively, these inherited diseases affect one out of every twenty souls.

Gaucher disease is one of genetic medicine's success stories: it is one on a short list (certainly compared with infectious maladies) of genetic diseases that are actionable today.[1] Gaucher disease serves as a model for how we may one day treat all genetic diseases in a hospital outpatient clinic as easily as we treat most bacterial infections.

Philippe Charles Ernest Gaucher was a dermatologist who practiced medicine in *fin de siècle* Paris. In that age of Baudelaire, Victor Hugo, and early modernism in the arts, Gaucher founded a new specialized medical journal on venereal diseases and was a professor of syphilography, the scientific study of the syphilis bacterium and disease. But today, Gaucher is remembered far more for a disease he was the first to describe as a medical student—in fact, it was the subject of his senior thesis.

Gaucher first identified what would become his eponymous disease in a thirty-two-year-old Parisian housekeeper who complained of pain in her left side. This discomfort was caused by an enlarged and tender spleen, an organ located in the abdomen, catty-corner to the appendix, where cells of the immune system gather to exchange information. There are many reasons why people's spleens become enlarged, including infections, cancers, and something called *idiopathic* causes. *Idiopathic* translates from medical-speak into contemporary American English roughly as "I don't know how the heck this patient got so sick." Also known in medicine as a "trash basket" diagnosis, the term is used when physicians are clueless about disease etiology but don't want to look dumb or unprepared in front of their patients. In this respect, doctors have perhaps learned something about obfuscation from their legal colleagues.

Given these options, the agreed upon diagnostic bin for this unfortunate young Parisienne was "cancer otherwise unspecified." However, Gaucher had seen many cancer patients on the wards, had attended their autopsies, and had looked at their cells under the microscope with the pathologists. There were no signs of cancer in the cells of her enlarged spleen, liver, blood, or anywhere else. This paradox is what made his medical student thesis thought provoking. Instead of cancer cells, Gaucher and his professors saw star-like, irregularly shaped white blood cells of the immune system called *macrophages*. These cells appeared as though engorged with micrometer-scale crumpled white tissue paper balls, with their center surrounded by a sort of halo. This was the characteristic finding of spleen macrophages in patients with Gaucher disease. These immune cells were unable to digest the refuse they had enveloped and swallowed. Three years later, a six-year-old French boy with similar looking cells in his spleen died, confirming Gaucher's original findings.

The enlarged spleens of the unfortunate young woman, the boy, and others subsequently identified as Gaucher disease victims remained idiopathic long after they died—in fact, until the 1960s. At that point, Roscoe Brady, a hardworking biochemist at the National Institutes of Health campus in Bethesda, Maryland, along with David Shapiro of the Weitzmann Institute in Israel and their colleagues identified a specific missing enzyme that caused the paper-like structures to accumulate in the macrophages of the spleen. Seven years later, they had purified enough of the missing enzyme from discarded human placentas to show that injecting it back into patients could help ameliorate this disease.

Thus was born enzyme replacement therapy for the first actionable disease to run in families, even before the gene had been identified.

In 1981, Edward Ginns and his coworkers, also at the National Institutes of Health, cloned the Gaucher disease gene and discovered the true molecular basis of this genetic malady whose mutation is carried (as a recessive disease along with another non-mutated copy) by about one of out of a hundred souls. Identification of the gene allowed industrial-scale production of the enzyme to help patients the world over—an early biotechnology industry success story.

Thanks to these unsung laboratory heroes, the Parisian housekeeper's long decayed spleen was picked out of the idiopathic wastebasket and transferred into a shiny new etiological hamper that didn't exist in Philippe Gaucher's era: genetic disease.

Many physicians, including me, have trouble pronouncing the name of the gene whose mutation causes Gaucher disease—*lysosomal glucocerebrocidase*—and so find it more convenient to use its nickname and official human gene symbol: GBA.

GBA is what molecular geneticists refer to as a housekeeping gene. The main domestic duty of GBA is to take out and dispose of a particular type of crumpled, tacky, and fatty cell trash, whose proper scientific name is glucosylceramide. When GBA isn't around to do the job, the glucosylceramide sticky stuff abbreviated as GL-1 builds up in the macrophages of spleen, which can eventually swell up to ten times its normal size. Depending on the particular patient and mutation, GBA also causes enlargement of the liver, as well as causing dysfunction in the cells of the bone marrow, skin, lung, and sometimes in its most serious form, the brain, all GBA-expressing places where macrophages patrol and mop up other cells' messes.

Almost a hundred years after Gaucher's discovery, Napoleon Jackson was born. He was raised and educated in one of the idyllic islands in the Caribbean archipelago. His extended family had lived there on the same family farm for generations longer than anyone could remember. There, he and his family experienced many tropical storms across the years.

As a young man, Napoleon moved to the New York City area looking for better opportunities and a chance to make it as a performer and musician. But he definitely did not move there for the frigid weather, and Napoleon never quite adjusted to the new climate. As a thirty-something, he had moved with his wife and children to New Jersey and, to support them, commuted to a job as a physical plant maintenance worker at one of the hospitals on the Upper East Side of Manhattan, an area nicknamed Bedpan Alley by local residents because of the density of medical institutions.

I saw Napoleon Jackson in clinic several days after another storm born in the tropics, Superstorm Sandy, hit the New York area. That terrible storm had hammered New York City and New Jersey with a ferocity I had never experienced. It was the largest Atlantic hurricane on record for the city and forever changed our landscape. Tens of thousands of trees were blown down. Whole communities were washed away, or, as happened on Staten Island, burned away by wind-whipped firestorms. Hundreds of residents were killed.

The hurricane had slammed my own house in Westchester, which lost electricity, heat, and nearly hope for eleven days. On the morning Napoleon walked into the clinic, I was feeling crumpled, chilled, and powerless after an epic commute into a post-Sandy, soggy, sans-subway city that looked like the scene of a Manhattan-gets-destroyed-yet-again disaster movie.

But my new patient was in far worse shape. Napoleon was exhausted, having barely having slept through the night for several days. His uniform was dirty, muddy, and disheveled. His pants were ripped. He had been through hell during the storm and was lucky to be alive.

He was suffering from a mysterious malady that left him exhausted. His bones ached. On examination, his primary care doctor had found that his spleen was enormous, so swollen that it might have ruptured, bleeding him out internally like a broken water balloon. This event could easily have transpired had Napoleon, say, slipped on the wet floor of the hospital basement where he had spent a long night during the hurricane, quietly but effectively saving lives.

When Sandy struck, just a short distance from the medical center clinic where Napoleon was visiting me, New York University's Langone Medical Center faced catastrophe when the hospital's basement, elevator shafts, and lower levels were swallowed by more than twelve feet of the swollen East River and its backup power generators failed. Almost the entire hospital lost power, including the surgical, medical, and neonatal intensive care units, where the sickest of sickest patients live and sometimes die. Patients, including premature infants on battery-depleted ventilators, were carried down nine flights of slippery stairs in the dark and outside into the fierce gale of the superstorm. Napoleon was one of many heroes who bravely struggled across the region all that tragic night long. The basement of the Manhattan hospital where he worked also flooded. Chilled and exhausted, he and all his coworkers fought the relentless Atlantic Ocean all night. Napoleon and compatriots risked their lives repeatedly, and successfully saved many people that night by preventing his hospital's patients from spiraling down the same East River toilet bowl as Langone Medical Center.

Napoleon's first symptoms of whatever it was that was ailing him had appeared years earlier, when routine blood tests showed low levels of platelets, the blood cells that prevent bleeding. Although this sometimes caused bruises, including tiny bruises called *petechiae* on the tips of his fingers when he played guitar, he shrugged it off and didn't let it interfere with his work or personal life.

Like Gaucher's original patient in Paris, Napoleon's spleen was filled with engorged macrophages. Because of low platelets and large spleen, Napoleon was worked up for possible leukemia, with his doctor ordering a bone marrow biopsy. The pathologist found none of the menacing-looking cell nuclei that would be characteristic of cancer cells, so it was not leukemia. The bone marrow did exhibit other atypical findings; the blood cells were more crinkled than normal and were filled with nanoscale bruises called *vacuoles*. However, while clearly abnormal, these findings were as murky as the East River in terms of achieving a diagnosis. Fortunately, since no cancer cells were found, he was never hammered with nonspecific cytotoxic chemotherapy. What was really happening was that the same fatty GBA was also filling up his bone marrow with macrophages, which caused him to become anemic and tire easily.

Next, a sleuthing hematologist following his low platelets noticed the fact that Napoleon's parents were distantly related because they grew up on the same island in the Caribbean. Islands have been called

"God's gift to geneticists" because they clue us in to the increased odds that we could sniff out a recessive genetic disease, which doesn't cause a problem unless a person inherits two copies of a mutated gene responsible for the disease, one copy from each parent. On islands, both real and figurative (such as a cultural island whose members tend to marry inside their own group, for example belonging to the same religion), the likelihood that carriers of the same recessive disease gene copy (we all carry two to five of such copies) will marry each another is higher than in a diverse community, such as a large city.

This meticulous hematologist hypothesized that Napoleon might have Hermansky-Pudlak syndrome, another rare recessive genetic disease where a different type of gene trash, this one more like wax paper with a different micrometer-scale appearance, builds up in the bone marrow. In addition, one of Napoleon's ancestors was an albino, with very light skin, blond hair, and pale eyes— features associated with Hermansky-Pudlak syndrome.

Despite that, Napoleon never had genetic testing for this syndrome because his particular medical insurance wouldn't cover the cost of the test. Why? Hermansky-Pudlak syndrome didn't meet the insurer's coverage criterion that it was an "actionable" finding. The company said that while the data might help inform him about his own future and that of his family, on a very practical level, testing wouldn't significantly affect Napoleon's own medical care.[2]

We appealed. I had what is called a peer-to-peer meeting with a doctor who worked for Napoleon's insurer to discuss his case. I presented his history and our findings, but the insurer refused us, finding a polite way to call me a pointy-headed academic who ordered too many tests for his patients—a Dr. Don Quixote chasing after diagnostic-testing windmills and bankrupting our medical system (though not specifically mentioning the cost to the insurance company). We tried unsuccessfully to find a research lab interested to perform testing of Napoleon's platelets with a technique called electron microscopy that could also make the diagnosis at no cost to him. Clinical research studies that offer no-cost genetic testing are often the exception to the rule that you never want your doctors to think you are an "interesting" case.

In addition to his spleen, Napoleon's liver was also enlarged. That, combined with the fact a brother and mother suffered from Parkinson disease, also put Gaucher disease high on the queue of Napoleon's most likely diagnoses. (Gaucher is believed to trigger Parkinson because those crumpled paper wads pile up in microglia, which are the macrophages of the central nervous system.) Napoleon was referred to a gastroenterologist who focused on his liver, which was so oversized that it reached across his abdomen, nearly touching his spleen, to try to help figure out the meaning of Napoleon's symptoms and signs. The liver doesn't rupture as easily as the more delicate spleen. That gave us a safer way to use any invasive procedures that might help us rule in or out possible diseases, a list also including another genetic disease, hemochromatosis, which causes iron to build up dangerously in the liver and other organs.

His gastroenterologist performed a liver biopsy. It was diagnostic coin flip. Coming up heads, we would find waxy, ceroid trash building up in the liver consistent with Hermansky-Pudlak disease. Tails, we would find white, foamy macrophages in the liver, indicating Gaucher disease. If the coin landed on its rim and was neither, we would have to start over again and re-think our way from the ground up.

Napoleon's liver cells under a microscope looked like they had been vandalized. Some cells were missing, replaced by patches of sticky, sugary, and protein filaments that are the body's equivalent of duct tape, which doctors refer to in more technical terms as *fibrosis*. The pathologist couldn't tell whether this had happened because of some toxin that Napoleon had been exposed to (industrial chemicals at work for example), or whether instead it was something his genome had done to him. Thus, these findings were relatively nonspecific.

Yet there was one particular finding the pathologist was quite emphatic about, and which pointed to a definitive diagnosis: the liver was loaded with precisely the kinds of foamy macrophages that are characteristic of Gaucher disease.

Because this was an actionable finding, his insurer agreed to let Dr. Don Quixote order genetic testing for Napoleon to confirm if it was indeed Gaucher disease. We drew his blood, and the lab analyzed Napoleon's GBA gene. The genomic sorcerer's apprentice that we call the *polymerase chain reaction* (PCR) was summoned to make millions of copies of his GBA gene. Developed in 1983 by Kary Mullis, Michael

Smith, and colleagues, PCR melts the DNA double helix's two strands with heat. As the temperature decreases, single-stranded primers (short synthetic pieces of DNA complementary to the specific region of the genome) search and find their sequence mates on the genome. Purified enzymes called DNA polymerases add new instances of A, C, G, and T to the tails of the primers, and the template sequence is replicated. These cycles repeat many times, and the newly synthesized PCR DNA pieces serve as templates as well, so that the desired sequence is exponentially amplified in a chain reaction like atomic fission.

In the next step, DNA sequencing, the process is repeated, with each A, C, G, and T now linked to techno-colored fluorescent probes that are excited by argon light, making short fragments. Finally, DNA sequencer parallel processors snap a gazillion nucleotide puzzle pieces together like an immense jigsaw, and we have our answer.

In Napoleon's case, we found the mutation we were looking for. He had two copies of precisely the mutation in the GBA protein, likely inherited through a distant ancestor common to both his mother and father, both of whose families were from the same Caribbean island. This single base change altered just one of 536 amino-acid building blocks that comprise the GBA protein, but that tiny change was enough to make Napoleon gravely ill.

The identical GBA mutation had been seen many times before. It is known to have first appeared about fifty generations ago during the early Middle Ages in a single individual who also happened to be Jewish. The origination of this mutation can be traced to around that time by analyzing the pattern of genetic variants surrounding it to see how often that surrounding pattern is present in different ethnic groups of people alive today. Each time a sperm and an egg form a child, the father's and mother's genomes recombine, or shuffle. By looking at the surrounding genetic variants and how much different ethnic groups share the particular combination, we can date the historical origin of individual mutations. Today, this specific single nucleotide mutation is a repeat offender: the Gaucher disease mutation observed all over the world, most often (but not always) in Jewish individuals.

When mutations arise in an individual who belongs to a specific ethnic group whose members tend to marry among themselves—such as

Jews—the particular mutation can become more common in that group than people outside the group. These are sometimes called *founder mutations*.

Jewish founder mutations in other genes, such as the breast cancer risk gene mutation *BRCA1* discussed in another chapter, are also observed in some individuals of Hispanic ancestry, possibly people who are the descendants of a single *Converso*. The Conversos were Jews forced to convert to Catholicism during the Spanish Inquisition in the sixteenth century to escape the *auto da fe*. Thus, although we will never know for sure, it is tempting to speculate that Napoleon had distant relatives who carried this founder GBA mutation and escaped to the Caribbean islands of the New World, unwittingly passing the mutation down through generations of increasingly distant cousins until two of them found each other and became proud parents of Napoleon.

Another possibility was that the identical mutation arose *de novo* in Napoleon's family, unrelated to any Converso, independently of what occurred centuries ago, and coincidentally happened to change the same GBA amino acid, leaving enough residual activity for the family members to survive.

The particular mutation carried by Napoleon partially destroys GBA's ability to do its housekeeping job. There has never been observed in humans or animals a GBA mutation that completely eliminates its activity and yet allows the mutation carrier to survive. Thus, it appears that complete loss of GBA is incompatible with life.

Now that we had a diagnosis, Napoleon started therapy. Treatment for Gaucher disease works is quite easy to understand for anyone who has had to manage a household. If GBA carries a mutation, and the protein it produces isn't completely able to keep up with its housekeeping, you get it an assistant to help—a molecular maid service. You inject people with a fit and able GBA protein that can dispose of the wrinkled spitball wads collecting up on the cellular floor.

To make GBA protein for Gaucher disease patients is biotech's equivalent of haute cuisine, with a bill approximating more than $400,000 for a year's supply. It is prepared as a kind of cell stew in huge vats. While usually made with animal cells, amazingly there is also a vegetarian version, with GBA made from carrots. When the Gaucher stew is done, it is strained and decanted so that only the GBA is left. If you only use the unmodified GBA, about 95 percent of the protein never gets to the macrophages that need it. The preparation coats the GBA with special sugars that macrophages have specific receptors for and very much like to eat, as well as some other choice ingredients, including some similar to the anti-oxidant preservatives and stabilizers that go into potato chips. Then, you inject it into the blood, where it is eaten up by "macs" in the spleen, bone marrow, and other organs that need it. Replacing just about 10 percent of the normal level of GBA is enough to do the job, as it is indeed a very hardworking protein.

Napoleon started getting injections of GBA enzyme replacement every two weeks—in this case, the kind made with the cloned gene in animal cells, not carrots. With an unambiguous genetic diagnosis and a medical literature trail documenting its benefits well, his insurer did not give him a hard time about starting the expensive enzyme replacement therapy. The molecular mechanistic maid service started to clean up the mess that had accumulated since his first birthday. After several months his macrophages had digested millions of the tiny foamy vacuoles cluttering up his bone marrow, liver, and spleen.

Napoleon was now getting better. His pain receded, and his blood counts amped up. Instead of being afflicted with a mysterious, chronic debilitating illness that could cause him to lose his life or livelihood—or (perhaps even worse) leave him frozen with fear about a future that could include multiple myeloma, Parkinson's disease, and pain—he was now little different from the thousands of others walking alongside him every day on their way into the hospital. Now he had more energy to play with his young children. His genetic risk of suddenly dying from spontaneously bleeding out was now hopefully just history. Gaucher therapy doesn't come cheap. Overall, each year about one-third of the total research costs to complete the Human Genome Project is spent on just this one drug for this single genetic disease. There are about ten thousand patients with Gaucher disease. About five thousand people around the world receive GBA enzyme replacement therapy, most of whom pay less than $400,000 a year and about 10 percent, mostly people outside the United States, pay little or nothing.

Why was his treatment so expensive? In part, exorbitant annual price tags larger than the one on a Maserati are the result of a US law passed in 1983 called the Orphan Drug Act.[3] This Reagan administration–era law tries to provide incentives to drug companies to work on rare diseases like Gaucher by giving them financial carrots, notably, the ability to charge pretty much whatever they want once a drug is approved for at least seven years (plus the inevitable delays from legal dodges that typically add several more years). The law prevents competing drugs for the same disease from being sold to treat the same disease, unless a competing drug can demonstrate superiority over the already approved one.

Without these incentives, the thinking goes, the pharmaceutical industry would not spend money on researching cures for rare disorders. It typically costs more than one billion dollars to undertake what is one of the longest, most challenging, and risky business endeavors in the twenty-first century: getting a drug to market in an over-bureaucratized, over-regulated, and over-litigious Western society.

There are now at least four companies making enzyme replacement therapy for Gaucher disease, although, curiously, the price hasn't come down much. Under our current terabyte of rules, regulations, and laws, precision genetic medicine doesn't come cheap. But failure to treat patients can also carry oversized costs. Left untreated, many Gaucher patients would be repeatedly hospitalized, surgicized, procedurized, and unable to work and benefit society in the many ways that people like Napoleon do.

Why are the actionable genetic diseases only a small minority today, despite the great genomic leaps forward we have made? For Gaucher and several other diseases that affect cellular "housekeeping" functions, a modified therapeutic product can be injected into the blood and reach the right place in the right type of cell to get the job done. Unfortunately, this isn't the case today for most genetic diseases. On the practical level, many of these are the "undruggable," meaning that we aren't (yet) clever enough to figure out how to precision guide the right molecule to the right place to fix the problem, at least without side effects that harm other organs along the way.

An important reason why many genetic diseases can't be effectively cured is that they are caused by mutations in many genes, not just one. Med-speak describes these maladies as "complex genetic traits," which refers to any medically measurable aspect of human beings that does not show the classic, simple patters of recessive or dominant inheritance that high school students are taught about genetics. There can be multiple genes whose mutations interact with each other in fiendishly complicated interactions to cause disease. Additionally, life experiences and environmental exposures can be part of the mix. Often, these diseases interest non-geneticists the most: obesity is a prime example of such a complex trait.

During my rotation as a medical student in psychiatry some years ago, an older physician confided to me that he would never hire obese staff members because he felt that they "lacked willpower." This psychodynamic view that obesity "is all in the mind" has now changed dramatically because it has been recognized as a group of diseases with identifiable physiological and neurological etiologies, and not just reflecting deficiencies in willpower or abnormal psychology.

Obesity, simply defined, is a collection of diseases with the common manifestation as excessive body fat, to the point that it may lead to shorter life expectancy due to increased health problems.[4] Medically, we define obesity as having a body mass index (body weight in kilograms divided by height in meters) greater than thirty. While the human gene pool has not changed very much in the past several decades, rates of obesity have dramatically increased in many countries during that period.

There are several rare "simple" single-gene mutations that have large effects and cause dramatic obesity. However, most obesity is thought to arise primarily from the interaction between hundreds of mutations in several dozen distinct genes,[5] the availability of different types of food, and levels of daily activity. Gene mutations can affect metabolic efficiency at which the body burns calories, the brain's ability to regulate appetite, efficiency of food absorption, psychological state, thyroid hormone levels, and many additional important biological mechanisms. There is even evidence that specific genetic mutations may affect preference for particular types of sustenance, such as fried food. We also have evidence that certain gene mutations repeatedly identified on several continents as influencing obesity in the current generation did not have the same effect on body mass and weight in our great grandparents' generation, when lifestyles were more active (walking instead of cars, farm work instead of desk jobs) and environmental exposures were different.

Several approved drugs can be prescribed for weight loss, but for most people, these drugs have a relatively modest impact and can be used only as adjunctive to diet and exercise.

Complex traits present a big dilemma for the drug industry. Our current approach to making therapeutics is analogous to big game hunters who go on safari with large rifles planning to shoot one big target, say, a tiger. These companies focus all their effort trying to go after large, obvious targets that, using current technology, can be shot and made into an annual five-billion-dollar-plus revenue trophy.

However, complex traits are a different type of beast. They are more like a swarm of bees. Shooting bullets at a swarm of bees, no matter how high precision the rifle, isn't going to do much good. Completely different pharmacological approaches to attacking complex genetic traits (something like smoke or insecticide versus bullets) are needed to make these undruggable diseases actionable, perhaps multiple drugs affecting multiple key bottlenecks at the same time, new technologies to regulate how genes are turned off and on, nano-robots patrolling the bloodstream, and many other innovations yet to be conceived.

Yes, the ability to alter complex genetic traits will definitely come, but we don't yet know when. Our advances in genomics makes me optimistic that the future will provide new illumination for diseases that today are the therapeutic equivalent of dark matter. A century ago the "French disease" seemed impossible to cure. Today, if we squint, an image comes into focus of a future when most genetic diseases, including complex traits, can routinely be cured.

TWO

The Unactionables

What you don't know can hurt you, and what you do know can also hurt you.

All animal species carry the burden of genetic diseases. One standard poodle I met in my neighborhood several years ago would straight out faint when the mail carrier appeared each day. This is because he, along with a pack of Doberman pinschers, Labrador retrievers, poodles, dachshunds, and other breeds, has canine narcolepsy, a recessive genetic disease that can make the dogs fall asleep at the drop of a table scrap or even mid-stride while running after a toy. The gene encodes a receptor, a type of antenna that sits on the outside of brain cells and responds to a hormone that affects sleep and arousal from slumber. Thousands of known genetic diseases beset animals, with likely many-fold more yet to be discovered.

However, while as a species we share with many animal species a common mechanism of genetic disease, namely DNA mutation, what uniquely distinguishes genetic diseases in *Homo sapiens* is, well, the human condition. Unlike, for example, our canine friends who largely can only live in the moment, when we are healthy and the sun is shining, we can envision that we still could catch the flu in the future and that rain could be on the way just over the horizon. We have consciousness, self-awareness, and the ability to comprehend that, one day, we can become ill and die. Our self-awareness allows us to anticipate the future and not have our consciousness boxed solely into the present moment. It's as if the human brain has a sort of double vision that allows us to observe the present while simultaneously trying to envision what will be happening in the future.

This ability to anticipate the future makes testing for the genetic diseases I will refer to as the *unactionables* so problematic. The unactionables are the antithesis of genes like Gaucher disease that one can test for with hope of an effective treatment and better days ahead.

The unactionables offer predictions of your medical future that many find troubling, or terrifying, to hear and, critically, cannot be changed: the risk of great suffering looming as destiny and for which there is (at least at present) no good treatment much less cure. These are the bêtes noires of genetic testing. Knowledge without power, the unactionables can be like Greek tragedy as symptoms inexorably unfold while everyone watches and anticipates.

Oddly enough, I met Lydia not as a patient in my clinic, but at a social event. We were both attending an evening salon in the district of Manhattan known as Chelsea, near the Hudson River. About twenty to thirty people had gathered at the flat of a couple, an academic and a financial professional, over a cornucopia of appetizers and an open bar to hear a visiting South African physicist discuss "What's New in Cosmology," a subject I knew nothing about, which was what attracted me in the first place. The visiting astrophysicist gave his lecture, which was followed by a vigorous discussion about predeterminism, the concept that all events are cosmologically determined in advance, and physical indeterminism, which says they are not (my limited understanding is that the latter view is more consistent with most of contemporary mainstream physics).

After the lecture and discussion, I was staring out from the twenty-first floor at the incredible starry sky of the West Village. While I was standing there at the windows admiring the view, Lydia walked up to me and introduced herself. She was a dark-haired forty-something financial professional who worked and lived in the city. Earlier that evening, before the lecture, I had been talking with another guest who had recently moved to New York from Argentina about what genetic testing is and is not useful for, and how people in different countries can look at the same information but interpret it very differently. For example, American women are more likely to choose having risk-reducing surgery after testing positive for mutations in the BRCA1 and BRCA2 breast-ovarian cancer genes than their counterparts in several European countries. Lydia had talked with the other guest, the subject had come up, and now she came over to me.

"Are you the kind of doctor who does genetic testing?" she asked. "I think my family needs this type of help."

The next week, Lydia followed up later by telephoning me at my office. She had called several times but left no message, often a sign that a patient is concerned about privacy and that the topic is sensitive. Now, recognizing this was the same local number, I picked up the phone.

"Hello, Doctor?" she began, reminding me of our original discussion. "I'm Lydia Gerstmann. We met at the lecture the other evening. My mother was diagnosed with ovarian cancer about two years ago. She died not long after turning fifty-seven. It was horrible watching her suffer, what she went through. I look a lot like my mother and we have very similar personalities. I'm very concerned that I may develop ovarian cancer like my mother. I've been reading up on this on the Internet, and if I had to, I would be willing to have surgery to prevent this. If my insurance won't cover this testing, I would still pay for it anyway."

This personal tragedy obviously had weighed heavily on Lydia's psyche, and it also was clear that she was worried that she might have inherited a similar fate. It also revealed a common general misunderstanding about genetic testing, that (except in the situation of identical twins and certain rare syndromes) a person who looks or acts more like a family member is more likely to develop the same disease, in this particular case ovarian cancer. In general, Lydia didn't seem to think that she was very much like her father and so wasn't as worried about inheriting his genes.

But as I took a family history over the telephone, it quickly became clear that there was no pattern that looked like any of the classic genetic causes of ovarian cancer in her family. There was no other close relative with ovarian, breast, uterine, or other cancer. Because most ovarian cancer occurs after the age of fifty, mostly without a known genetic etiology, and she was not in a higher risk ethnic group, overall there were no compelling reasons based on what I had heard to recommend her pursuing this further and coming in to clinic in order to be tested. For example, her risk of having a mutation in the notorious BRCA1 or BRCA2 breast-ovarian cancer risk genes was estimated to be around 3 percent.

However, there's an old clinical pearl (the sort of useful tip you give to medical students) in medical genetics: if you are called to see a child with six fingers, the last thing you look at are the hands. That way you don't become so distracted by the obvious problem that you forget about everything else that might be going on. In this way, medical genetics is an opposite of emergency room medicine, where you are usually forced urgently to focus on the acute problem at hand and deal with other issues later.

When I questioned Lydia about non-cancer-related diseases in her family, she stated that her father had developed Alzheimer disease in his early fifties.

"Early fifties? Are you sure it's Alzheimer disease?" I asked.

"Yes," she said. She was sure. "He was cared for at St. Francis Medical Center, in New Jersey. I can get you his doctor's name if you like. But the doctor never said anything about this being genetic."

Familial Alzheimer disease is an unmentionable unactionable that is among the most dreaded of maladies. A tremendous amount of outstanding literature has been written on this topic, and so I will only very briefly summarize here. Alzheimer disease is the most common form of dementia, affecting more than four million Americans, and is the sixth leading cause of death in the country. It is, of course, also a scourge around the globe as well. Almost all of its victims are older than sixty-five.

Alzheimer disease itself typically begins with subtle failures of memory. You can't remember where you parked your car, or whether you walked the dog earlier this evening. As the fog becomes thicker, Alzheimer victims can become deeply confused, agitated, and socially withdrawn. In many instances, the rest of the body can remain in relatively good health for years while the neurological and psychiatric symptoms progressively worsen. However, eventually, Alzheimer disease can become incapacitating and death ensues, often from general lack of concern about oneself and one's surrounding, particularly lack of interest in eating and personal hygiene. A typical course lasts eight years, but this can range from less than a year to more than twenty years. Alzheimer disease is the only one among the ten most common causes of death in America that currently has no widely accepted medical interventions to cure, slow, or prevent it.

It is increasingly being recognized that Alzheimer disease has at least three stages. In the first stage, individuals have specific, definable molecular alterations that in many cases are detectable before symptoms have developed. In the case of genetic mutations, these are detectable many decades before the first signs of memory loss. Proposed nongenetic pre-symptomatic markers of disease in this first stage range from brain imaging and blood biomarkers to software that analyzes subtle changes in the content of people's e-mail and digital phone conversations. In the second phase, the disease has subsequently progressed so that clear symptoms appear (such as memory loss), and in the third, advanced phase the disease inexorably continues toward its conclusion.

However, most of the first stage pre-symptomatic biomarkers, including the results of genetic testing, don't provide highly deterministic predictions of a person's future. For example, much has been written about the genetic test for a gene called APOE e4. The relationship between this gene mutation and Alzheimer disease, typically occurring after age sixty-five, is well established in many ethnic groups, particularly for those who carry two copies of this mutation. Precisely how it causes early onset Alzheimer disease is still somewhat confusing and a matter of debate. But notably, almost half of people diagnosed with Alzheimer disease do not carry even one copy of this mutation. This genetic test is not generally considered at present in clinical practice to be useful in identifying individuals pre-symptomatically in the first stage of Alzheimer disease, although it is likely important for research studies working to discover new preventive and ameliorating therapies. A mutation in another gene, TREM2, has also recently been identified that increases risk of developing Alzheimer disease after age sixty-five to about one in four. Like APOE e4 testing, the precise place for TREM2 genetic testing in pre-symptomatic Alzheimer disease in those without a family history, along with several other candidate genes being investigated in various stages of the research pipeline, is still a subject of debate in the medical community.

In contrast, genetic testing does indeed have a more accepted role in the medical community for people who have family members or close relatives with early-onset Alzheimer disease. About one-quarter of Alzheimer disease occurs in kindreds where two or more close relatives have been diagnosed. Additionally, somewhere between one in fifty to one in twenty Alzheimer patients are called early onset because they develop clear symptoms before the age of sixty-five. Familial history and early age of onset together are considered reasonable evidence to consider genetic testing.

Children or siblings of a person afflicted with Alzheimer disease have an estimated lifetime risk of one in four to one in five of developing the disease, which is about two and a half times more than the average person. Having more than one affected close relative on the same side of the family further increases risk, in particular when one of them develops Alzheimer disease before age seventy, to a risk of about one in three to almost one in two.

With early onset Alzheimer disease, when a close family member develops dementia before age sixty-five, three genes are currently known to cause a much higher chance of developing symptoms. All of these three genes, Presenilin 1, Presenilin 2, and Amyloid Beta A4, play roles in the processing of beta-amyloid, a kind of protein trash that can't properly be disposed of and so backs up in brain cells like dishwater in a clogged sink (similar to the last chapter's Gaucher disease, but this time in the brain instead of liver, spleen, and bone marrow). Mutations in these genes confer much higher probability of developing the disease during one's lifetime to something like nine out of ten people, or more.

Lydia remembered well how her father first became forgetful, like losing his wallet. With time, he had trouble keeping up with his job as a government employee in New Jersey. He saw a neurologist and had a battery of tests. After being clinically diagnosed with Alzheimer disease, she said, her father gradually became "a shadow of his former self" over several years. At first, her mother quit her job and took care of him full time. Later, he was cared for in a nursing facility as he became more and more agitated and confused. During this stressful period, Lydia's mother unfortunately developed ovarian cancer and subsequently passed away from her disease.

Lydia then mentioned that her uncle Martin, also on her father's side, had been diagnosed with Alzheimer disease in his later fifties as well and died from it several years later. Because her uncle had married a woman from Arizona and moved to that state, she didn't have as much detail about him, but she was confident in his diagnosis.

The rest of the history on the grandparents, great-grandparents, cousins, and such was somewhat spotty. Lydia had a brother, but he was estranged. Small families can present an obstacle to establishing a significant family medical history, but that was what we had to work with. Despite the missing family history details, what I had heard was worrisome. This was in the ballpark for familial early onset Alzheimer disease.

Another troubling fact was that while her surname was German, her father's side of the family had come to the United States from Central Russia. This rang a bell. Familial Alzheimer disease is more common in the Volga Germans. This was an ethnic enclave encouraged to immigrate to Russia by Catherine the Great in the eighteenth century. Perhaps because they were culturally somewhat distinct from the surrounding Slav ethnic groups, the Volga Germans often tended to intermarry, and thus became a *genetic island*. As I explained earlier, this term refers to a group of people that intermarries among themselves rather than marrying outside the group, often for cultural reasons. Since Alzheimer is an adult disease that doesn't become symptomatic until ages after people are done having kids, it can be passed on easily. In this way, a late onset disease like a high-risk familial Alzheimer disease gene mutation could spread through many Volga German families and make the precise DNA alteration from the founder become more common in that ethnic group. In summary, her ethnic background increased the overall risk that her family could carry one of the high-risk familial Alzheimer disease "founder" mutations, and hence increased my level of concern.

"Can you get copies of your father's and your uncle's medical records?" I asked.

"If you think it's important I can." She paused. "What about my mother's records? So you don't think I will get ovarian cancer?"

"From what I am hearing, you don't meet any of the present guidelines we usually use that would suggest you need genetic testing for hereditary ovarian or breast cancer. We can still do this, but in general, insurance companies won't pay for the testing."

"Oh, thank you. That's great news," Lydia said. She paused. She described herself as a "do-er" who liked to be proactive. Lydia wanted to know if there anything she could still do to make the cancer even less likely. I suggested she exercise regularly, try to eat a healthy diet, keep her weight down, and give up smoking completely (which, of course, I recommend to all, whether they are considering having genetic testing or not).

I paused. "However, I would like Connie, a genetic counselor who works with me, to call you to help collect some more information and medical records on your family."

I continued, "I'm actually kind of concerned about what you told me regarding your dad and uncle having Alzheimer disease. I would like Connie to talk with you and find out about getting some of their health records.

"If we can confirm from their medical records that your dad and uncle had Alzheimer disease, then I'd like to have you see a neurologist for a careful examination. After that, we may want to discuss testing, or a research study, for Alzheimer disease risk genes."

Lydia paused, caught off guard. "When I called I wasn't really thinking about this. I don't know very much about Alzheimer disease testing. How much does it cost?" She now seemed more concerned about the expense than she had earlier, when her concern was cancer. "And this is a blood test or a, uh . . . brain test? A brain test sounds bad."

"It's only a blood test," I reassured her. "We can talk with your insurance company about whether or not they'll pay for testing. However, there are also some other issues that would be better to go over in person, and so that's why I'd like you to come in."

Again, Lydia paused. "Well, I can see how surgery can help with something like breast and ovarian cancer, but isn't the brain different? So what can you do about Alzheimer disease? Are there drugs you can take to prevent it?"

"No, there aren't any medicines you can take," I said. "There are some early studies that exercising regularly, eating a healthy diet, keeping your weight down, and giving up smoking can help."

"But you told me to do those things already," Lydia said. "Exercise more, eat healthy and better. I know I should quit smoking anyway, and I guess this is as good a reason as there is. But besides that, if there's nothing I can do to help stop becoming senile like my father, and I had the gene, it would just make me upset and anxious all the time." She paused. "So what is the purpose of having the test if there isn't anything you can do about it?"

"Well, some want to know in order to plan their lives accordingly. They may want to change their career, or focus on what's most important for them, or use the information to help ensure that their children don't inherit the mutation."

"You can use the testing to help prevent having kids who have an Alzheimer disease mutation?" Now she was intrigued. "Well, I did freeze my eggs several years ago when I was younger, just to play it safe if I did decide to have kids later on.[1] So this could be useful, but I'm not planning to have children right now. Well, I have to go now. Thank you for your time, Doctor. Let me think about it, and I'll get back to you and Connie."

And so, the very same person, Lydia, who was very eager to pursue genetic testing for an actionable, ovarian cancer risk, and very prepared to act on it if need be, was much less enthusiastic about pursuing genetic testing for the unactionable early onset Alzheimer disease, where currently there is no clear action to take if she tested positive. As a postscript, to this day Lydia has not pursued Alzheimer disease genetic testing, at least not at my clinic.

There is a debate in the scholarly literature as to whether the potential for depression, anxiety, and other types of distress including, most notably, the risk of suicide, should preclude the clinical use of genetic testing for unactionables like Alzheimer disease. Several years ago, a well-designed randomized clinical trial of APOE e4 testing for asymptomatic adult children of parents who had Alzheimer disease showed that participants who chose to have genetic testing and receive the results had not exhibited significantly more anxiety, depression, or distress prior to testing compared with those who chose not to be tested.[2] However, not surprisingly, once the results were available, those who tested positive had significantly higher measures of distress than those who tested negative.

All clinical trials have limitations to their conclusions, and one important limitation was that (for obvious ethical reasons), potential participants were excluded if at the beginning of the study they admitted to considering the option of suicide if tests results were positive for the APOE e4 gene. But that suggests that the very people who were potentially most vulnerable to the worst adverse events from the genetic-testing results may have been excluded from the trial's analyses, complicating how this testing would play out when tested on potentially hundreds of thousands of patients. In 2013, Dena S. Davis, an ethicist at Pennsylvania's Lehigh University, raised a related, and certainly controversial, issue in the debate over genetic testing for unactionable Alzheimer disease: whether suicide should be considered as a rational actionable preventative strategy for healthy people when carrying a high-risk mutation for Alzheimer disease or other forms of dementia.[3] This troubling and complex viewpoint has brought the medical community to a place it has never been before in the debate about euthanasia.

A strategy for how to handle some of the tangled complexity in testing for early onset Alzheimer disease has planned, to some degree, by medicine's experience with testing for Huntington disease. HD is characterized by progressive and increasingly severe mental and physical deterioration that can often begin when a patient is still in the prime of life—thirty-five to forty-five years old—HD by definition afflicts its victims early.

In 1983, scientists announced that they had been able to determine that the HD mutation is in a gene on the short arm of human chromosome 4. After a decade of intensive research, the 1993 identification of the actual HD gene was hailed as "the crown jewel of recent neurogenic discoveries" by the then-director of the National Institute for Neurological Disorders and Stroke (one of the US National Institutes of Health).[4]

Early testing for the HD mutation had actually begun before that gene itself was located. By identifying and then tracing markers linked to the gene in various family members, researchers at Massachusetts General Hospital, Johns Hopkins, and their collaborators spanning as far as Venezuela developed a test in 1986 that was initially about 95 percent accurate, with diagnostic acuity improving somewhat over the next few years as more testing centers got involved and more genetic markers were identified.

As the testing accelerated in subsequent years, a collaborative international model for testing and related genetic counseling began to evolve, eventually leading to guidelines developed by the World Federation of Neurology and the International Huntington Association. The guidelines outline a multifaceted approach to counseling patients, a process designed to begin well before any testing actually occurs and involving a large interdisciplinary team of experts ranging from medical geneticists and a genetic counselors to neurologists, psychologists, and social workers.

The guidelines recommend an initial pre-screening interview, which can be done over the telephone, followed by three more extensive in-person sessions that cover genetic counseling as well as neurological and psychological evaluations. If an individual decides to proceed with testing, results would be disclosed at a fourth session, with follow-up sessions over the next two years.[5]

In terms of whether testing leads to large numbers of suicides among people who are at risk but pre-symptomatic for developing HD, a Canadian study from the 1990s suggests that it does not. By then, among over 4,500 people tested for the HD mutation in twenty-one countries by then, five had committed suicide and another twenty-one had attempted it. Of the five who did commit suicide, however, two had reached the point that they were already showing signs of the disease. For context, suicide rates among those who have developed HD symptoms are already as much as ten times higher than they are for the general US population, because HD can directly cause depression.[6]

Perhaps in line with Lydia's personal decisions regarding early onset familial Alzheimer disease, relatively few of those eligible to get testing for HD actually do: only about 15 percent.[7] By contrast, the screening rate for inherited cancer mutations can run as high as 95 percent. These findings amount to a strong hint that the low rate of HD testing is largely a factor of this frightening disease's status as an unactionable unmentionable, as opposed to cancer syndromes that become actionable through, say, targeted screening for early-stage tumors or pre-cancerous conditions (for example for breast or colorectal cancer). While testing for each genetic disease has its distinct idiosyncrasies and nuances, it is tempting to speculate that the situation is similar for familial Alzheimer disease. Presumably, screening rates for such presently unactionable disorders as Huntington and early onset Alzheimer disease would be much higher if and when more effective preventative strategies were developed. In the meantime, people who do choose testing cite a desire to plan their lives accordingly and maximize control of their destiny.

Hippocrates of Kos is among the most famous physicians in recorded history. He lived during the Age of Pericles in ancient Greece, but his influence is still very much alive today in the practice of medicine. Physicians still recite a modernized version of the Hippocratic oath. While they leave out the original oath's appeals to the ancient gods Apollo and Hygieia, the ethical imperatives that constitute this oath still guide modern medicine, almost three thousand years later. The overarching ethical principle that should guide any medical, drug, surgery, or diagnostic procedure (including genetic tests) is how it balances risks against benefits. When doctors discuss any action with patients, they should do so in the context of balancing risks against benefits on the hypothetical Scales of Hippocrates. With regard to the unactionables, there is a huge range of opinion, and not enough hard data, about how to weigh the benefits and risks of genetic testing differentially for these diseases. Clearly, the best way to improve the situation with unactionable genetic tests such as for Huntington and early onset Alzheimer disease is to have effective therapies and strategies to prevent the disorders associated with them. However, until then, the interest in testing for the unactionables will be likely be limited to those who are interested in the knowledge for its own sake, those who wish to plan their future lives accordingly, and those who don't want to pass on these mutations to their children. We are unlikely to see any sustained consensus for presymptomatic testing of people for mutations in unactionables without strong family history anytime soon.

THREE

Altitude Sickness

Paraganglioma is a great word to know if you are playing Scrabble with your family around the dining room table, netting a whopping fourteen points, not including double or triple letter scores. But paraganglioma is not at all a winning word if you know it is due to your family cancer history.

Paragangliomas arise from nervous system cells perched upon large blood vessels running down the body midline from where the head meets the neck to the bottom of the buttock cheeks. They are like sentries in a middle guard tower whose job it is to detect oxygen. When these cells do not sense oxygen, they release chemical signals into the blood to speed up both heartbeat and breathing, in order to increase blood flow, blood pressure, and oxygenation in the process. This is part of the fight of the "fight or flight" response.[1] Many of these cells can gather together in a pack as a paraganglioma, raising a chemical ruckus that causes palpitations, sweating, panic attacks and, in extreme cases, even heart attacks. More ominously, the paraganglioma cell pack itself can morph into a cluster of cancer cells, take off down those same blood vessels, and run wild around the body, wreaking havoc as metastases.

Paragangliomas are a very rare type of tumor in people, but also occur in cows, pigs, and even African elephants. Of course, all animals need oxygen. Because they have the same basic genes coding for the same basic hardware as do we *Homo sapiens*, the same genetic diseases generally occur in many different species of animals.

Rosalind Temasek grew up as one of eight kids in a large family in the Rocky Mountains. Her childhood was healthy and about as medically unexciting as one could wish for. A scholar and athlete, she had come "back East" after college to New York City for a great professional opportunity in a Fortune 500 company based in the skyscraper canyons of midtown Manhattan. Here, she had met her husband, started a family, and juggled all the myriad responsibilities that come with the two-career urban family lifestyle.

On a beautiful fall New York City day, when probably the last thing a busy executive and even busier parent wants to do is go see a doctor, Rosalind taxied across town to tell me her story. She was in her thirties. Tall, poised, and wearing executive casual, Rosalind was expecting her second child in about three months. But a complication had arisen.

Rosalind's mom had two cousins by blood. Both of her cousins developed paraganglioma tumors in their thirties. The unusual and challenging-to-articulate name for a diagnosis highlighted the surprising nature of this coincidence for the families. Indeed, fewer than five hundred of these cancers are diagnosed each year in the entire United States, and so the odds of this happening merely by chance to relatives living not far away from each other at similar ages are similar to those of running into two celebrities randomly walking on the same city block at the same moment.

This unusual coincidence had prompted the family's surgeons to recommend consultations with clinical geneticists. About one-quarter to one-half of paraganglioma diagnoses occur in people affected by genetic diseases. There are fewer than two thousand medical geneticist physicians in the entire country (with post–medical school residency training positions only half occupied, so this situation unfortunately has the potential to trend further downward in the future), and they are disproportionally located in major metropolitan areas. So it was fortunate that their doctors were familiar with hereditary paraganglioma syndromes and were able to refer family members for genetic testing, even though the first family members to develop paragangliomas did not live near a large city.

Even though it is a very rare tumor, there are at least nine genes (and possibly more) whose mutation causes paragangliomas and its only slightly less rare cousin, the pheochromocytoma. The para-pheo genes are active during the first few months of pregnancy, when stem cells destined to create the brain and nervous system arise in what is called the *embryonic neural crest*. From the neural crest, some of these neuronal stem cells migrate during fetal development in order to help the brain regulate blood pressure. The cells destined for paraganglia climb, jump, and wrap around arteries and veins, while their cousins the chromaffin cells migrate to the top of the kidney and form the adrenal gland (*ad-renal* in Latin means literally on top of the kidneys). In response to danger, the adrenal gland can quickly release chemicals called *catecholamines* into the general blood circulation that help instigate the "fight or flight" response. Pheochromocytomas are sometimes called *adrenal paragangliomas*. These cells are, in fact, almost impossible to distinguish from paragangliomas when viewed through a powerful microscope, although it's much easier to tell one type from the other if you can see the surrounding tissue in order to map their origin.

Why mutations in so many genes cause such a rare cancer is not clear but perhaps reflects the fact that cells of the nervous system tend to express (that is, make RNA and protein) more genes than cells in just about any other organ. With more moving parts, more things can go wrong.

Some of the other genes related to familial paraganglioma are unusual in that they can only cause disease if inherited from the father. This is because they have the atypical property that they are only expressed from the paternal copy of the gene, through a process called *imprinting*.

The vast majority of genes "turn on" and make RNA and, after that, protein from both the mother's and father's chromosomes in the genome simultaneously. However, a little less than two thousand genes are referred to as *imprinted*, meaning a gene is expressed and makes RNA and then protein from only one parent's copy. Some imprinted genes are expressed only from the mother, and some only from the father.

Imprinting was originally discovered in the early 1980s, when experiments involving the transplantation of male or female nuclei (containing all genomic DNA) into one-cell stage embryos confirmed that both maternal and paternal genomes are important for a fetus to develop normally into a baby who is born healthy. Despite the wishes of some who might believe they are perfect the way they are, we can't reproduce with ourselves but need to find another person as a mate in order to have a child. The imprinting mechanism marks one copy of the gene with chemical modifications called *methyl groups*, which act like molecular nametags marking what is from Dad and what is from Mom so their contributions can be clearly distinguished and kept track of properly.

There is an aspect of parental rivalry in all this. Imprinting is thought to be important for genes that play particularly important roles in promoting cell proliferation during pregnancy, perhaps also related to keeping up the right oxygen levels in the womb for what is perhaps the ultimate battle of the sexes in evolutionary parental conflict. This idea proposes that mother and father each have different investments and interests in the fitness of their respective genes in a child. Thus, three paraganglioma risk genes (with the hard-to-pronounce names SDHD, SDHAF2, and, ironically for some parents with sons, MAX) are turned off by the mother through the imprinting mechanism. Since this trio of genes is only expressed if inherited from the father, if mutated they only cause paragangliomas if inherited from his side of the family.

Rosalind's two cousins had their blood drawn and their DNA sequenced, and several months later were told that each of them carries the identical mutation in one of the familial paraganglioma genes. As predicted by Rosalind's family history, this was not one of the imprinted genes causing paragangliomas, but one of the non-imprinted genes that caused problems in both men and women. The particular gene in question here is called *Succinate Dehydrogenase B* (SDHB).

Overall, more than ten people in the family were tested. Rosalind's mother was one of those, and during the process of screening the many relatives, learned to her surprise that she carried the same mutation, although into her fifties she had never had any noticeable related health issues. However, another cousin, Martin, who also had the mutation developed gradual onset of palpitations, headaches, and anxiety attacks in his early twenties. This turned out to be caused by not one, not two, but *three* previously undiagnosed paragangliomas riding around the top of his kidneys. The majority of paragangliomas are benign in that they are never able to metastasize to other parts of the body, infiltrate organs, and cause problems. So in this case, despite the multiple opportunities for trouble and seemingly against the odds, all three were successfully surgically removed and Martin cured.

Ironically, these paroxysmal attacks, unpleasant as they were, may have saved Martin's life, a biochemical silver lining. If the tumors hadn't barked up a storm, triggering all those anxiety symptoms, they might have gone unnoticed until they accumulated further mutations and eventually metastasized. This is particularly a problem for SDHB, which of the nine genes is most likely to cause the tumors to migrate and disseminate. At that point, these tumors become extremely difficult to cure. (One genetic counselor told me her mnemonic to keep straight all the genes that cause this syndrome was that the B in SDHB stood for "the Bad one to get.")

The precise triggers from genes and environmental exposures that cause one person to develop disease and another with the exact same mutation not to still leave us scratching our heads. Mitochondria are the motors in cells that breathe oxygen and use it to make the energy needed to run the body, such as when the body needs to run away from a predatory animal. When one of two copies of the SDHB gene is mutated, the mito motors tend once in a while to sputter and spin their gears. This leads to the production of oxygen radicals that can damage DNA and cause mutations to be spit out of the mitochondria.

SDHB also helps to break down a particular chemical called *succinate*. In general, succinate normally builds up when cells aren't getting enough oxygen. When succinate accumulates, the cells feel like they are constantly gasping for air. The succinate builds up and sticks to certain proteins, decorating them and changing their shape. Like a baseball centerfielder wearing a parka in the outfield on sweltering day in July, the succinate-decorated proteins just can't get the job done as well as they should. The succinate molecules are sticky and puffy, and bump into other proteins, like a shape-shifting row of falling dominos.

This causes a kind of intracellular panic attack called the "cell stress response," which informs the body that more cells need to be made to complete their oxygen-sniffing job. This causes the stressed cells both to divide and to make catecholamine hormones that transmit a chemical alert that more oxygen is needed. While the same mutation is present in all cells of the body, the paraganglion cells rely on the proteins in the mitochondria that get stuck to succinate particularly acutely to sniff oxygen and so are affected more than any other type of worker cell.

According to one research hypothesis, living in the lower oxygen levels at higher altitudes might contribute to paraganglioma cancer risk. So perhaps it is more than coincidence that Rosalind's family grew up in the relatively thin air of the Western Rocky Mountains. In fact, recent imaging studies suggest that paragangliomas may be at least five times more common than we thought among people who live at higher altitudes, and presumably, pigs, cows, and elephants who live on mountain plateaus as well. But few of these paragangliomas reach a size that causes any discernable medical problems, and so they remain undetected.

I think that everyone can appreciate how altitude and the quality of the air we breathe can influence the course of different diseases. Historically, perhaps the most well-known example is the use of mountain sanitariums for patients suffering from tuberculosis before the development of effective antibiotics against this disease. In the novel *The Magic Mountain*, by German author Thomas Mann (whose tubercular wife repaired to the five-thousand-foot elevation Alpine Waldsanatorium, in Davos, Switzerland), as well as in Italian composer Giacomo Puccini's operas *La Bohème* and *Tosca*, and in many other works in the era preceding medicinal chemistry, tuberculosis was known as the malady *consumption*.

The classic sign of consumption was coughed-up blood. Coughing also helped the disease spread. The *Mycobacterium tuberculosis* bacillus was most often transmitted from person to person by tiny droplets of water coughed out of the lungs and into the air. To treat this disease without modern chemical and genetic tools, physicians in the pre–World War I era used to recommend "chasing the cure." This meant that patients with consumption should seek high altitude abodes with rarefied clean air, where their disease would be fought with what was ironically at the time called "managed care." In this era, there were hundreds of sanatoriums in the mountains of Europe, and hundreds more in the American West. The town of Colorado Springs, not too far from where Rosalind's family lived, was essentially built on the back of turn-of-the-century-era sanitariums. The city as a whole was marketed by its Chamber of Commerce as "America's Greatest Sanitarium" (the terms *sanatorium* and *sanitarium* were often used interchangeably) and included facilities that ranged from little more than tent cities to one plush retreat for the wealthy that featured the amenities of a luxury cruise ship and even printed its own literary journal. For several decades, treating TB was the growing city's major industry, with thousands employed by the sanatoriums.

The mechanisms whereby tuberculosis bacteria grow more slowly at higher altitudes are not completely understood. The tuberculosis bacterium uses oxygen and grows (slightly) more slowly at higher altitudes, thereby consuming its host's health and body less when the amount of oxygen in the air is reduced at rarefied heights. Other potential causes of this phenomenon include the lower oxygen level's effect on the lung's acid-based pH balance, or even the higher UV skin exposure at altitudes where the air is thinner. Simply getting away from the polluted, industrial-era air of crowded cities likely helped some patients survive. Still, it's probable that any perceived cure rate was simply a result of the fact that, according to modern estimates, perhaps as many as 30 percent of patients could survive the disease even without antibiotics.[2]

It's possible that the cure wasn't even as minimally effective as it appeared to be. One of the main side effects of chasing the cure was altitude sickness. Before the widespread availability of cars and airplanes, long journeys, followed by early days of altitude sickness in a sanatorium, may have helped select for healthier patients. After the discovery of anti-tuberculosis antibiotics, which were far more effective, higher-altitude managed-care therapy largely fell by the wayside, and the sanatoriums were closed or scrubbed clean and converted into spas.

Because of how altitude might aggravate her disease, and contrary to most of my thinking, I found myself in the unusual situation of telling Rosalind that despite the traffic exhausts, smoking, and factory air pollution, it was possible that moving to New York, which is at sea level, had been good for her health (presuming she stays out of the street when daredevil taxi drivers zoom by). Similarly, familial paraganglioma mutation carriers are strongly urged not to smoke. Smoking tobacco or vaping e-cigarettes not only reduces oxygen levels in the lung but also acclimates the O_2 sensors in the brain to tolerate lower oxygen levels. Needless to say, it is important also to avoid secondhand smoke (which, fortunately for her and others, is now outlawed in New York City restaurants and stores).

The best offense is a good defense. To keep Rosalind doing all that she accomplishes every day as a successful working mom, she needs to try to avoid the kinds of stress that could lead to paroxysms of palpitations, sweat, panic, and the like, the same symptoms that bedeviled her cousin. This is tricky, of course, because we all have these fits every occasionally: when, for example, your boss and the babysitter call at precisely the same moment you are battling heavy traffic. But particularly when the paroxysmal panic attacks erupt spontaneously without obvious precipitating factors, we do tests to see if the levels of the paraganglioma chemicals are abnormally high.

Another patient, Terri, developed anxiety in her mid-thirties that started with a trot and progressed to a gallop over the course of about two and a half years. She had a recurrent feeling on being on edge—her whole body felt tense and clenched, even during such humdrum tasks as vacuuming the living room. Her hands would begin to tremble like an elderly woman's. Her heart would race and her temples pound with each heartbeat. She would become short of breath, her face would turn white, and she'd almost pass out. To relieve her anxiety she would jog, bike, or run on the treadmill.

In this case, exercise turned out to be good for her health in an unanticipated and indirect way. Terri developed right hip pain from running on pavement, an occupational hazard for Manhattan joggers. This led her to get a sports medicine MRI of her hip. In the corner of the MRI of her right hip, an astute radiologist at our hospital picked up an incidental finding: a light gray smudge on top of her right kidney. That smudge turned out to be the cousin of the paraganglioma, the pheochromocytoma. Terri had been convinced that her panic attacks were purely psychosomatic, the normative sequelae of modern urban married life in the age of anxiety. Many people who have these tumors feel stigmatized because they are told it's all in their head.

While, on the one hand, having a pheochromocytoma is not a reason to be happy, on the other Terri was relieved to learn of this identifiable physiological source of her anxiety, and an actionable one at that. After her surgery, she recalls waking up and, from that moment on, feeling relieved and being able to relax for the first time in years. Terri has done well since then and has had no additional paragangliomas or pheochromocytomas identified.

In patients with paraganglioma and pheochromocytoma syndromes, we recommend continued screening for paraganglioma tumors even without symptoms, in order to catch them early. Similar to some of the patients with hereditary breast cancers, we examine the patients and do MRI scans annually to pick up suspicious lumps and bumps. The MRI scans pretty much everything except the head, arms, and legs.

Rosalind found out that she carried the SDHB mutation when she was twenty weeks pregnant. Highly educated and with a complex job in corporate finance, she understood that sophisticated genetic reproductive technologies could tell whether her unborn child was at risk for familial paraganglioma. Today, it is possible to test the fetus for this mutation just from drawing the mother's blood, without having to perform any invasive tests involving long needles inserted into the womb or other invasive procedures.

For those wishing to have children, there are a number of advanced reproductive technologies (ART) available to the ever-increasing number of families and extended families with one of the approximately three to four thousand genes linked to diseases.

The essence of ART is the incredible ability to combine sperm and egg submerged in special liquid media in a dish outside the body. ART developed during a technological revolution that led to culturing human cells outside the body in the 1960s and 1970s. In ART, sperm and ova are mixed together overnight in a dish housed in an incubator set at body temperature and the right oxygen levels to simulate a uterus. The next day, fertilization is confirmed by observing a single cell nucleus that has divided into two. This verifies that the single cell called a zygote is on its way to becoming a fetus. The cells continue to divide. By five to six days after in vitro fertilization, there are now hundreds of cells that form a hollow ball. On one side are the cells destined to become the fetus, and on the other pole are cells that will become the placenta, formed by the fertilized zygote to help support the embryo until birth.

The first pregnancy and live birth after the fertilization of a human egg outside the body was more than thirty years ago, in 1978. Since then, more than four million in vitro (in glass) healthy and successful pregnancies have been achieved worldwide. Early predictions that ART babies by the year 2000 would outnumber "traditional" naturally conceived children and replace sex were perhaps too optimistic, or pessimistic depending on your point of view. However, at least one prominent geneticist, Lee Silver of Princeton University, has publically articulated "a future in which people will not use sex to reproduce. [With the availability of prenatal genetic testing], that's a very dangerous thing to do."[3]

ART pregnancies currently account for about 2 percent of live births in the United States and Europe. Initially, there was anxiety about the unknowns of increased pregnancy complications and long-term risks to children born of ART pregnancies. Indeed, some early studies suggested tentative associations between ART and medical problems in children. The majority of these studies used relatively small numbers of children. As recently as about a decade ago, some studies proposed that ART could potentially cause higher rates of problems with fetal imprinting (described earlier in this chapter) that cause expression of growth-related genes from both parents' chromosomes, and thereby cause child cancer and overgrowth syndromes.

At the end of many of these articles reporting the results of research in the medical literature, the authors often include phrases such as "further studies are needed to confirm these findings more broadly." Unfortunately, popular news media like television sometimes will pick up on these initial studies, broadcasting them as established fact, when they really only preliminary. The nature of statistical analysis of scientific and medical data is such that there inevitably will be initial claims based on small-scale studies that are not further substantiated in larger and more diverse groups of people (or are merely so-called false positives). The initial, smaller studies are important, in that they can act like canaries in a coal mine, helping to insure that bona fide problems (true positives) are identified early.

The laws of mathematics and statistics make it is essentially impossible to perfectly avoid false signals in some early, limited studies. As a society, we can set the bar so that we take on fewer false positive studies, but then we run the risk of missing the true positives. The necessary evil false positives cropping up in smaller medical studies is one of the reasons governmental agencies like the Food and Drug Administration often have to intervene to help assess the totality of the evidence base (or lack thereof) and whether regulatory interventions such as outright bans on procedures or medicines are required.

Now that several decades of children born from ART pregnancies can be evaluated, the more robust statistics of larger numbers show that there is no strong evidence that higher rates of any imprinting disorders are associated with ART. However, with regard to longer-term effects, such as adult cancers, diabetes, autoimmune disorders, or other disorders, particularly for more rare conditions, we may have to wait.

An embryo at risk for inheriting genetic disease can be evaluated by a process known as *pre-implantation genetic diagnosis* (PGD). Through "assisted hatching," a technician can prick the outer surface of an in vitro fertilized embryo with a laser, chemicals, or even a finely drawn glass needle. Two to three dozen cells can then be gently taken from the cells destined to become the placenta through a hatched opening on its outer surface, and the genetic material sequenced to look for a specific mutation. Embryos that undergo this process largely survive no worse for wear, but do carry a higher likelihood of becoming identical twins. Embryos found by the genetic tests to carry harmful mutations are not implanted.

Since all technologies are imperfect, the error rate of the current PGD technologies for recessive genetic diseases (two mutated copies of the gene and no un-mutated copies) where a bona fide mutation is missed is about two per hundred, while for dominant genetic diseases (one mutated copy, one un-mutated copy) is about one in ten.

In the next decade, we are likely to see a new generation of pre-implantation genetic diagnosis, which could lead to the possibility of actually repairing genes in ART embryos affected by genetic disorders. Recent powerful basic science advances using a technique called *CRISPR* (pronounced "crisper," and the acronym for the tongue twister "clustered regularly interspaced short palindromic repeats") have offered robust evidence that it's possible to repair gene defects in embryos of several mammalian model organisms, including nonhuman primates. This is an inexpensive, remarkably effective gene-editing technique easy enough to be performed in literally thousands of laboratories around the world. The technique enables editing of DNA sequence changes that can be curative for both recessive and dominant genetic diseases. It can also be used in patient-derived embryonic stem cells, which can then be expanded in the lab to help ameliorate genetic diseases for such hard-to-treat areas as brain and heart tissue in patients living now, as well as to alter an individual patient's sperm and egg to reduce the burden of genetic disease in future generations.[4]

CRISPR is a bacterial immune system gene that protects a bacterium against invading viruses. Also called *Cas9*, it was originally discovered in 1987. Because its purpose was not well understood, Cas9 was initially thought to be unimportant bacterial DNA without any function. Now we know that these genes can be deployed in a way that makes it possible to actually "edit" genes in humans and other species. CRISPR can cut both strands of DNA and can be co-injected into fertilized embryos with a guide RNA that can be modified to alter most of the changes possible in the genome. It has been used to cure mice with genetic diseases involving the liver. Chinese scientists have created monkeys with artificial mutations by using CRISPR to modify otherwise normal embryos, including mutations in three different genes involving the immune system and diabetes. Once the monkeys were born, the CRISPR-introduced changes were present in most, although usually not all, of the monkeys' cells, because the CRISPR protein typically doesn't actually perform its surgery until after the one-cell embryo has divided more than once. Genome sequencing of the monkeys after their birth did not yet reveal any "off-target" unintended mutations.

Subsequently, Chinese geneticists in Guangzhou published the first study of CRISPR gene editing on human embryos to correct beta-globin mutations that cause the disease beta-thalassemia. In this case, however, they worked only with embryos that would never have been able to be brought to term and born after pregnancy in a surrogate mother and would otherwise have been destroyed. Regardless, their work showed that the present CRISPR technology did not work as well in human embryos as expected. For many embryos, CRISPR did not correct the beta-globin mutation. More concerning, by using whole-genome sequencing, they discovered "off-target" new mutations introduced at other sites, including the delta-hemoglobin gene, which has DNA sequences similar to those of beta-hemoglobin. All in all, the effects of these off-target mutations if the embryos had been brought to term are difficult to predict.

Because of the technique's potential for powerful gene editing, many scientists, including its inventors, have pleaded for a global moratorium on its use until physicians, scientists, governments, and the public understand more fully any accompanying risks.

While clearly a great deal more work needs to be done, this technology is quite well developed already. I anticipate that (likely initially from outside the United States) in the next decade we will hear about gene-edited children born with mutations successfully corrected by CRISPR. As discussed above, as with ART and PGD, longer-term safety issues and effects of the CRISPR gene-editing process will hang over us, likely for decades.

For mothers who have ART/PGD procedures, several complications can arise. The procedure is invasive: the ova have to be removed by ultrasound-guided aspiration in order to be fertilized in vitro. While perhaps not as bad as many medical interventions, for some women who psychologically consider themselves healthy and not "patients," this can be a deal breaker. Another more important issue is how the parents feel about potentially having more than one child from the resulting pregnancy. Since several fertilized eggs are generally implanted into the uterus, and each one can become a child, multi-child pregnancies are common with ART; two women have given birth to eight siblings (octuplets).

Additionally, there is the issue of high cost, something not likely to be solved in the United States. Several rounds of ART can easily run into tens of thousands of dollars, some of which may be covered by insurance. Pre-implantation genetic diagnosis typically adds three to five thousand dollars to this total when performed in the United States, although the advertised cost is less in Asian medical centers.

These procedures can also lead to complications. About two out of a hundred ART cycles to induce pregnancy lead to hospitalizations because of such complications as *ovarian hyperstimulation syndrome* (OHSS). Mild OHSS occurs in about a quarter of ART pregnancies. This causes abdominal pain from ovarian enlargement triggered by the hormones that are injected to induce ovulation, a general inflammatory response with increased white blood cells, bloating, cramping, and nausea. Severe OHSS occurs in about one in every thousand ART cycles (each woman can have five, six, or more cycles to achieve a successful pregnancy). The high hormone levels cause the liver to make proteins called fibrinogens that can trigger the formation of small and large blood clots at different body sites. While rare, this can potentially be catastrophic. OHSS can cause the kidneys and lungs to fail, and in extreme instances, can lead to death for both mother and child. Some researchers and physicians have suggested that the use of high doses of reproductive hormones might increase a mother's risk of breast, endometrial, and ovarian cancer, but this is complicated by the fact that certain combinations of reproductive hormones can actually be protective, while others increase risk. Additionally, female reproductive hormones may actually reduce the incidence of other malignancies, such as colon cancer. Studies so far, over several decades now, have not revealed any clear link between ART and a mother's risk of cancer.

Overall, these problems, inconveniences, and costs can raise many doubts for parents who are carriers of genetic disease mutations and might be interested in using ART and PGD. Yet, when asked, between 25 percent and 75 percent of couples with high-risk genetic diseases say they would prefer trying the technologies to the lower-tech options: taking their chances naturally and pursuing abortion if the child turns out to carry the disease mutation. The large dispersion in different studies on interest levels in ART and PGD (about half the couples) perhaps speaks to the very personal and subjective nature of this decision.

ART and PGD have a powerful appeal. These technologies can empower parents with the ability to do what many want more than anything else in the world: better the quality of the lives of their children, their children's children, and their children's children's children, in significant ways.

Consider the case of Dakshine and Rama Venatharam.[5] They were twenty-three and twenty-four years old, respectively, and had been married for two years when they had their first child, Vanani, a daughter with a recessive genetic disease called *beta-thalassemia major*. This disorder is caused by mutations in the beta-globin gene, one of the two main components of hemoglobin. Hemoglobin is one of the most highly produced proteins in the entire human genome. It binds directly to oxygen. Red blood cells, which carry oxygen throughout the body, are essentially packs of hemoglobin, having lost even their cell nucleus and DNA throughout the course of evolutionary history so they can fit in more hemoglobin bound to oxygen. Vanani had two copies of a single-base pair mutation in beta-globin, one of the two proteins that combine to make hemoglobin. Since there are more than two hundred known beta-globin mutations, it was likely that her parents were related by blood through a shared ancestor.

Patients with beta-thalassemia major, like Vanani, have severe anemia, meaning they cannot get enough oxygen to the body's different tissues. They need repeated blood transfusions to increase the number of functioning red blood cells that will ferry oxygen to its needed destinations, often about once a month. Since correctly functioning hemoglobin also uses iron to help catch and release oxygen in the lungs and different body parts, the transfusions result in iron buildup in the liver, bone marrow, spleen, and other organs, which causes dysfunction in these organs. While patients can be further treated with iron-binding drugs known as *chelators* to remove the iron from the body, children with beta-thalassemia major generally fall off the growth charts and have many medical problems. Overall, quality of life is significantly hurt, and life expectancy is typically about twenty-eight years. For people of Indian ethnicity like Dakshine and Rama, about three out of every hundred persons carry one mutated copy of the beta-hemoglobin gene, so having children with beta-thalassemia is common.

Dakshine and Rama wanted to have a larger family. Their next two pregnancies also had beta-thalassemia major. Having undergone the emotionally difficult experience of caring for a child with beta-thalassemia major, they terminated each of these pregnancies. For their next child, they sought to use PGD and ART. Dakshine had two IVF-PGD cycles. In the first cycle, four oocytes each were retrieved successfully. Of these, two were not mature and so not usable. One of the two successfully fertilized eggs turned out to have beta-thalassemia major (a pair of mutated genes, one from each of the two parents), and so was not used. The other fertilized egg didn't result in a pregnancy. Since the first cycle of IVF wasn't successful, a second was pursued five months later. This round generated four IVF fertilized embryos. PGD testing revealed that each carried one (dominant) normal and one (recessive) mutant beta-globin gene, meaning that the disease couldn't inflict a child from that embryo. Three embryos were transferred, and this resulted in a successful pregnancy with one child.

There is strong consensus that the application of pre-implantation genetic diagnosis for pediatric genetic diseases that cause significant suffering or early death should be widely available, and more than three-quarters of parents facing this situation express interest in exploring its application.

There is less of a consensus on the use of ART and PGD for adult-onset genetic diseases, such as familial paraganglioma syndrome, carrier status for recessive diseases, or even the less severe forms of beta-thalassemia, where life expectancy can reach into the sixth decade of life.

The Ethics Committee of the American Society for Reproductive Medicine (ASRM), whose members provide these services, concluded in 2013 that "PGD for adult-onset conditions is ethically acceptable as a matter of reproductive liberty."[6] The precise number of parents having ART/PGD performed for adult-onset conditions is not well known, and this procedure is not regulated in the United States at the state or federal level.

In the United States, Europe, and Japan, PGD for adult genetic disorders is widely available. There is currently great variance of opinion among the general public, different medical specialties, and individual medical providers about whether the potential risks of PGD outweigh the benefits of preventing a serious disease that a person may or may not develop over a lifetime.

Arguments offered in support of PGD for serious adult-onset conditions include the right to parental reproductive choice, the medical good of preventing the transmission of genetic disorders, and potential social benefits of reducing the overall burden of disease. Parents may also wish to avoid the angst that they are not being good parents and are not doing all they can to better their children's lives. For their own peace of mind, couples may also want to do whatever they can to reduce the uncertainty and fear that may shadow their sons' and daughters' childhood years if they know they are likely to be affected with serious adult-onset conditions. Couples may also desire to avoid the "survivor's guilt" some children might experience knowing that they will not develop genetic diseases that affect their brothers or sisters.[7] The American Academy of Pediatrics, on the other hand, currently recommends that genetic testing for adult-onset conditions for which interventions are unavailable is generally inappropriate until children reach adulthood.[8]

In some cases, parents have different thresholds for themselves and their children: for example, some children whose parents have Huntington disease do not want to know their own mutation status for personal or insurance reasons, but do want ART and PGD testing when they have their own children (that is, grandchildren of individuals who have developed Huntington disease and children of a parent at 50 percent risk of inheriting the gene mutation) to prevent passing on a mutation—just in case.[9] This can be an ethically difficult situation for medical providers when these ART/PGD patients are found *not* to carry a mutation, since these patients really don't need ART/PGD, but the provider can't reveal that information to them as per their wishes.

In many situations, there are currently available treatments to ameliorate, but not cure, the genetic diseases for which ART/PGD is performed to prevent passing them on to the next generation. For example, frequent colonoscopy and/or surgery are used in thousands of individuals with Lynch syndrome to identify and prevent the gastrointestinal cancers estimated to occur in about seven out of ten Lynch syndrome mutation carriers who live to age seventy. Aspirin can be used for chemoprevention to reduce this risk further. Furthermore, even while the risk of developing gastrointestinal cancers is greater in Lynch syndrome carriers, those who do develop the cancers have a better prognosis than those who develop the same cancers from other causes, such as the natural vicissitudes of aging and accumulating sporadic mutations. The next several decades will likely see greatly improved imaging technologies for cancer surveillance and new drugs for prevention and treatment. And complete genetic cure might well become commonplace using the aforementioned gene editing technologies. So there is much debate about whether ART/PGD should be used for these types of adult genetic diseases where carriers may in fact never develop the disease.

The main argument against ART/PGD for adult genetic diseases is that problems are not always certain to arise in the future, a reality that must be balanced against the costs and risks of ART and PGD today. There's also the potential for technical misdiagnosis of embryos. A more speculative concern: this could because a slippery slope, leading to more frivolous uses of the same genetic technologies, including the potential selection for height, athletic prowess, or intelligence.[10] The idea of using of technologies such as ART and PGD to make "designer babies" may sound like the stuff of futuristic fiction. But it might be worthwhile to consider the use of one variation of pre-implantation screening that is less black and white than a case of serious pediatric genetic diseases: gender selection of children.

For millennia, couples have tried to influence whether they have sons or daughters, a desire that spawned myriad superstitions. These include having intercourse facing toward the North or South poles (presumably related to differential fluctuations of magnetic fields on Y-chromosome-carrying sperm motility), making love during different phases of the moon, and timing intercourse based on alterations in a woman's body temperature. There is (as advertised on television) even a diet that supposedly increases the probability of having a daughter by increasing dietary calcium and magnesium. None of these seem to be particularly effective.

Ancient cultures, including both those of Greece and Rome, often employed infanticide as a means of gender selection, usually by simply exposing newborns to the elements. From one translated Roman papyrus comes this directive: "I am still in Alexandria. . . . I beg and plead with you to take care of our little child, and as soon as we receive wages, I will send them to you. In the meantime, if (good fortune to you!) you give birth, if it is a boy, let it live; if it is a girl, expose it."[11]

King Henry VIII's famous attempts to select a male heir led, in part, to the creation of the Church of England, separating the nation from the Roman Catholic Church and its ban on divorce. As the newly installed Supreme Head of the Church of England, Henry was able to annul his marriage to Catherine of Aragon, who had not been able to bear

Henry a male heir. Bio-archaeologist Catrina Banks Whitley and anthropologist Kyra Kramer speculated, in a 2010 article, that Henry suffered a general difficulty to sire children who could survive past early infancy with what turned out to be six wives because of a genetic anomaly: his genome expressed a rare "Kell" antigen in his blood group.[12] That protein triggers immune responses in reproductive partners whose blood do not carry this variant against the fetus. A mother and father with conflicting Kell variants can produce a healthy first baby, but with any subsequent conceptions with the same father, the mother's immune system will tend to attack the fetus, often leading to miscarriage or neonatal death. The only way to know if the authors got this right would be to locate and exhume Henry's remains and hope to find for enough DNA in, say, bone, to sequence.

Today, ART and PGD enable child gender selection with high accuracy. Driving parents to select their children's sex are reasons ranging all the way from avoiding X-chromosome-linked genetic diseases—diseases that affect males and females differently (e.g., Fragile X syndrome, causing mental retardation in boys)—to personal preferences for family gender balance or having sibling companionship with children of the same gender. In some cultures, a preference for male children is very strong.

In the European Union, the task force on ethics for the leading professional body on this topic, the European Society of Human Reproduction (ESHRE), in 2013 recommended accepting PGD for gender selection to extend beyond sex-linked genetic diseases, while leaving some room for interpretation, such as in cases of families that have children only of one gender.[13] The American Society for Reproductive Medicine (ASRM), a leading professional society for this medical specialty, initially rejected PGD for child gender selection, but later revised its position to allow PGD sex selection for purposes of family balance of children's gender and parental preferences for use of child-gender selection, demonstrating the lack of consensus on this issue.[14]

However, in the United States, no federal laws currently exclude the use of PGD for gender selection of a child. Estimates are that about four out of ten US-based fertility clinics report offering ART/PGD for nonmedical child-gender selection. Some ART/PGD clinics publically advertise PGD for child-gender selection, as do similar clinics abroad that cater to American and European "medical tourism." Conversely, because some countries ban PGD for nonmedical child-gender selection (such as the People's Republic of China, Germany, United Kingdom, and Canada), the United States is a medical tourism destination for this purpose as well. One prominent center that offers such service, the Fertility Institutes, advertises on its website that it has PGD centers in New York, Los Angeles, and Guadalajara, Mexico, and has a future facility in India ("*Coming Soon!*") and states forthrightly: "If you want to be certain your next child will be the gender you are hoping for then no other method comes close to PGD (Preimplantation Genetic Diagnosis.) While traditional sperm-screening techniques have a success rates [*sic*] of 60–70%, only PGD offers virtually 100% accuracy."[15]

In Israel, while gender selection is not generally promoted, a couple with multiple children of one gender can apply to the national Ministry of Health for an exception to the rule; the ministry's approval would allow the nation's universal-health-care system to pay for ART and PGD for gender selection.

Even without ART and PGD, there other alternatives. Abortion is one of the most common strategies used by couples worldwide when faced with the difficult situation of having a child carrying a genetic disease. But this approach is highly dependent on cultural mores and relevant laws.

On one side of the spectrum, in the People's Republic of China, physicians can order an abortion (for genetic or nongenetic reasons) even if the parents don't want to have one. On the other end, countries that do not allow abortions even if the child has a genetic mutation causing a likely highly morbid or fatal disorder include Vatican City, Malta, and Iran.

On this issue, with an emphasis on individual autonomy and free choice, the United States and European Union countries are in the middle, and for thousands of genetic diseases, abortion is a common treatment used by many couples when they discover that their child is at risk.

The use of abortions to select the gender of a child is currently available in forty-two United States, except Arizona, Illinois, Oklahoma, Kansas, North Carolina, Pennsylvania, North Dakota, and South Dakota. As writer Emily Bazelon noted in *Slate* in 2014, "South Dakota goes so far as to require doctors to ask women seeking abortion if they know the sex of the 'unborn child' and are having an abortion because of it. In Oklahoma, your husband or parent or sibling can sue you if you have an abortion for this reason."[16]

In a majority of cases, PGD is performed to increase the chance of having boys and more often involves first-born children. In the United States, four out of five couples seeking PGD sex selection want boys. These decisions often appear to be driven by preferences by male heirs in some of our ethnic sub-cultures and are tied to inheritance rights, financial support for elderly relatives during retirement, carrying on the family name, and gender-specific religious roles.

In the People's Republic of China, ART and PGD for child gender selection is unambiguously illegal. Nevertheless, data suggest that this law is clearly not enforced, likely in response to a history of culturally driven gender selection, in which a pattern of aborting female fetuses has led to gender ratios above 117 boys born for every 100 females.

India's experience in the use of reproductive technologies for gender selection, including but not limited to ART and PGD, is particularly notable. Many powerful cultural forces in India, particularly among majority Hindus, lead to strong preferences for having sons. An important influence is the widespread dowry system. When a daughter is wedded to her groom, this ancient custom stipulates the bride's family transfer wealth to the newlyweds, in the form of a dowry, which can amount to six years of income.[17] The result, according to a saying, is that "raising a girl is like watering a plant in your neighbor's yard." There are other cultural preferences as well for sons over daughters, including the cultural imperative that sons financially support elderly parents. For India's Hindus, who represent about 80 percent of the nation's population,[18] religious principles also promote sons, but not daughters, lighting the funeral pyre of the parents to secure salvation after death.[19]

In the 1970s, amniocentesis and a more invasive technique called *chorionic villus sampling*, both of which can distinguish male from female embryos, were introduced into India. Soon they were being actively promoted for use in prenatal gender selection by the Indian medical profession.[20] Prenatal sex selection was seen as helping India to meet its population-control goals by stopping families from reproducing until they had a son, as well as by reducing the practice of female infanticide. A 1977 Indian study of couples who were early adopters of amniocentesis for gender selection described a 96 percent abortion rate for prenatal sexed girls and a zero percent rate for boys. After the introduction of fetal ultrasound in the 1980s, similar findings were confirmed in larger case studies. The Indian census of 1991 showed skewing of the gender ratio, and in 1994, a law banning the use of prenatal technologies for gender selection made their application illegal. Aspects of this ban were appealed, and in 2003 the Indian Supreme Court specifically banned the use of ART and PGD for gender selection of unborn children.[21]

However, technology is mercurial and adaptable. Since the mid-2000s, it has been possible to assess the gender of an unborn child with only a finger prick blood test of the mother during the seventh week of pregnancy. These tests look for male Y chromosome cell-free fetal DNA in the mother's blood. While no Indian company makes these home-use kits, American companies can sell these tests for home use and have done so since at least 2006. It appears that such test kits are finding their way to Northern India, where the gender ratio there has reached almost two boys for every girl born (the worldwide average is about 105 boys born for every 100 girls; a ratio that demographers speculate is nature's way of balancing the fact that both male infants and adults tend to have more fatal health problems, and that male adults tend to kill each other more).

Bucking the trend toward preferring males, surveys show that parents in Spain and Iceland slightly prefer female over male children, according to an international Gallup poll on children.[22] Whether Spain and Iceland have common or different motivations that lead them to the same preference is not clear, but in part this may reflect a preference for larger families. In the United States, about four out of ten couples surveyed have no preference. Among those who do have a preference, sons are favored by about six out of ten couples.

I have encountered parents interested in gender-selection specifically for diseases that affect males and females differentially, with especially keen interest for PGD among families who already have children with autism spectrum disorder.

Autism is significantly more prevalent among boys than among girls, a trend especially notable in some of the most severely affected children, such as those with highly repetitive behaviors. Girls appear to be better protected from both familial and *de novo* (e.g., for those with no family history of the disease) autism spectrum disorder symptoms. This disease is a genetically complex trait, with hundreds of different genetic variants that can contribute to its language problems, stereotyped behaviors such as hand flapping, learning disabilities, and other manifestations. Yet other factors, such as the male hormone testosterone and its derivatives, appear to modulate the expressivity of autism in complex ways, perhaps by altering specific regions of the brain during development. It can be a highly devastating disease for both children and the entire family.

If parents together have one or more children with diagnoses of autism, their chances of having another autistic child are elevated. Thus, even though autism is very rarely caused by a single genetic alteration, the use of ART and PGD for gender selection in favor of girls appears, at least anecdotally, to be on the rise.

According to the World Health Organization, the United States spends about 17 percent of its entire economy, three trillion dollars in 2013, on health care.[23] There are almost a million physicians and several million more allied health professionals. In academia, government agencies, private sector companies, not-for-profit organizations, and hundreds of policy lobbying organizations, there are thousands of health policy experts weighing in with their opinions on the role of genetics in prenatal medicine, along with many more would-be experts. However, at the end of the day, medical decision making remains largely in the domain of personal preference and autonomy for the individual patient. Patient autonomy should trump precision genetic medicine at all levels, and, I hope, will remain that way.

On the matter of whether to apply powerful existing and new emerging technologies in genetics to their unborn child, Rosalind and her husband didn't hesitate for a moment. ART, PGD, and abortion were medically available in New York State, financially within reach, and intellectually and emotionally accessible to them after genetic counseling. However, they chose to go "old school" for this most personal decision for them, their two-year-old daughter and unborn child. While perhaps this belies current stereotypes of high-powered New York power couples, Rosalind was from a large family from the American West with strong traditional family values. According to Rosalind's thinking, she, her parents, cousins, aunts, uncles, and all the rest had survived much worse. Her family had lived through depressions and recessions, natural disasters, and personal tragedies that transcended the genetic problems they had encountered. The family had thrived even if they did have the "bad" SDHB gene mutation that caused panic attacks and cancer diagnoses. Rosalind's husband, also from a large and successful family with strong traditional values, was of the same mind. He stood by Rosalind and exuded confidence that they would persevere. Four months later, after an uneventful pregnancy, they had a beautiful baby daughter. Rosalind is doing well and her family is thriving.

FOUR

The Blindfolded Poker Player and the Smoked Salmon

Michael X. Braedenberger was referred to me through Rick, a mutual friend we both knew from our undergraduate years at Princeton. Michael was an interesting guy, Rick told me, a brilliant self-made quantitative hedge fund manager. If I Googled him, Rick said, I'd get a dozen pages of hits—articles about his business in the *Wall Street Journal* and *Businessweek*, about his philanthropic work, as well as society gossip in blogs from Chicago, where he lived.

He'd also been diagnosed several years earlier with a genetic disease called the *MAP polyposis syndrome*. Rick told me he had recently had his genome sequenced and had a lot of questions about what was found beyond what his genetic counselor told him.

Rick also told me that Michael was obsessed with privacy. He had stored the data about his genome on a website with such rigorous security features that even the National Security Agency wouldn't be able to hack it. Soon enough, I'd experience for myself Michael's strong penchant for privacy, in the form of an "unlisted" Manhattan restaurant designed for privacy and discreet conversation. The restaurant had an unlisted telephone number. Its entrance was on the side, an unmarked red door, while the front looked like it was under construction. The white-cloth covered tables were spaced far apart in a nineteenth-century coach house décor.

Our bodies have at least seven different ways that every day protect the genome against the many and diverse insults that constantly attack it. These continual threats emanate from the very air we breathe, from the ultraviolet and cosmic rays from the sun and other celestial bodies, as well as to the inevitable accidental genetic DNA mis-incorporations, mistakes, and miscues that accumulate as part of the normal aging process as cells divide daily, repeatedly, over many decades of our lives.

Earth's first known forms of life, types of DNA-carrying bacteria, arose more than three billion years ago. At the time, there was almost no oxygen in our planet's air and these early bacteria breathed other gases, such as methane, to survive.

Fast forward another billion years, give or take a few hundred million, and cyanobacteria evolved that could perform the incredible feat of using the energy from the sun's rays to help them grow and make energy via photosynthesis. Ancient cyanobacteria were epically fruitful, multiplied ferociously, and took over all the warm and sunny real estate on the planet.

However, in doing so, cyanobacteria precipitated what is referred to as the *oxygen catastrophe*. Photosynthesis, by which plants use the energy from light to sop up carbon dioxide from the atmosphere and release oxygen (the dioxide in carbon dioxide) as a waste product, caused oxygen to accumulate in the planet's atmosphere. Oxygen is a highly unstable, toxic chemical that can mightily damage DNA and RNA, as well as many other biologically critical macromolecules like proteins and lipids. Oxygen flings electrons at these other molecules at the speed of light violently knocking off whole pieces or breaking molecules apart entirely.

The evolutionary success of cyanobacteria and the rise in atmospheric oxygen at the time thus caused what is thought to be one of the most significant mass extinction events in earth's history, wiping out almost all other forms of life at the time because of two factors. First, bacterial life at the time was not adapted to this terrible mutagen and could not repair the oxidative damage fast enough. Additionally, oxygen reacted with methane in the air to cause a reverse greenhouse effect, precipitating a drop of several degrees Celsius in the Earth's average atmospheric temperature. The results were similarly disastrous: the longest ice age in our four-and-a-half-billion-year history.

Fast forward several hundred million years, and equilibrium is reached. Today, oxygen is about 21 percent of the air we breathe, and we cannot survive without it. If oxygen goes away, so do we, rather quickly and unpleasantly. Like a campfire needs oxygen to burn the wood logs feeding it, our lungs and bodies use the awesome destructive power of oxygen to break down molecules storing energy and provide us with the fuel to do everything that living beings must to stay alive. In parallel, our immune system has even successfully weaponized toxic oxygen to our advantage. Specially constructed patrolling armored immune cells called *macrophages* and *granulocytes* use oxygen radicals as a weapon to kill invading infectious organisms and even cancer cells.

To protect themselves against this terrible atmospheric pollutant we call oxygen, bacteria, plants, insects, and later humans developed a kind of particular restorative pathway called *DNA base excision repair*, which focuses on fixing damage from the oxygen-caused mutations that constantly arise in our bodies.

From their humble beginnings on ancient earth to the age of humans, these genes have also evolved themselves to add on new ways to protect the genome, like nucleotide Swiss army knives.

In addition to oxygen, there are other chemicals that cause similar damage to DNA, for example charred foods containing oxidized carbon, nitrogen, and sulfur containing macromolecules. The bacteria that live in our gut, called the microbiome, and the immune system that polices these "good" bacteria by producing their own oxygen radicals as natural antibiotics when needed, also produce what is called *oxidative stress*. The lower intestine has particularly high levels of sulfur (which you likely already know, since most of us are unlucky enough not to have *mercaptan anosmia*, the genetic condition in which you cannot smell flatulence), which like oxygen can radicalize and damage a person's normal gut issue. This can be particularly problematic for the cells that line the gastrointestinal tract. There, of the more than one hundred different types of cells that make up our bodies, those lining the gastrointestinal tract divide the most rapidly, turning over every few days. That means by the time most people have their first screening for colorectal cancer at age fifty, the stem cells lining their colon will have divided more than one hundred and thirty thousand times. The more times a cell divides, the more opportunities for oxidative or sulfur stress to cause damage when the DNA is copied, which in turn increases cancer risk. More recently, other types of sulfur-producing long and skinny *Fusobacteria* have been shown independently by scientists across different continents to increase colorectal cancer risk, likely by also increasing inflammation and stimulating the colon cells to divide additional times.

There are also additional genes important for repairing DNA damaged from environmental chemicals. In families that have passed down certain mutations in the DNA repair gene through many generations, family members were once prone to develop cancers arising from the cells that line the stomach. But that trend changed. After the widespread introduction of refrigerators to store food in the early to mid-twentieth century, descendants carrying the very same mutations developed stomach cancer far less often. One explanation: people used to preserve meat for long periods by using techniques such as smoking and curing. Smoking kills any infectious bacteria lurking in the meat that can cause us to get sick, but also generates chemical compounds from the process of combustion. Incorporated into the smoked meats, these compounds can damage DNA after they are eaten and broken down in the stomach.[1] These include oxidized and alkylated proteins, fats, and other chemicals known to cause chemical modifications of DNA resulting in mutations, as well as other types of potential carcinogens that have been associated with increased cancer risk.

Smoking, curing, and pickling traditionally used soaking in water with different salts to brine different cuts of beef, poultry, pork, and fish. While brining would make the meats tenderer, it was also potentially problematic because some traditional recipes would use nitrate salts. These increase formation of noxious chemicals, including one type called *nitrosamines* that can glom onto DNA and confuse polymerases about whether to insert an A, C, G, or T base pair during cell division. (Nitrosamines are also found in tobacco smoke and e-cigarette vapors.)

The US Food and Drug Administration limits nitrite levels used in meat and meat byproducts because of concern over the potential to increase cancer rates in the general population,[2] a problem certainly not limited to families with DNA repair related genetic syndromes. More recently, the World Health Organization based largely on epidemiology findings, but not more conclusive clinical-trial data, concluded that eating processed meats like bacon, ham, and hot dogs can modestly increase risk of colon and stomach cancer.[3]

However, smoking, curing, and pickling recipes for preparing food, which have been around for hundreds and perhaps thousands of years, have become embedded as part of our culture and heritage.

We also know that inflammation is an important risk factor for several gastrointestinal tract cancers. Inflammatory bowel disease is a complex genetic and environmentally provoked disease that causes chronic irritation of the cells lining the intestine and colon and is known to increase the rates of colorectal cancer. Additionally, the other genome, the microbiome of bacteria that populate the gut, influences cancer risk. Chronic infection by the bacterium *H. pylori* and the inflammation that ensues trying to evict it from the stomach is a well-established provocateur for gastric cancer, in part because of collateral damage from the immune system's weaponization of oxygen. Generally, *H. pylori* is easily eradicated with several weeks of the right antibiotics and drugs that modulate the acidity of the stomach.

In developing nations with limited public health infrastructure, stomach colonization with *H. pylori* can be almost ubiquitous and contributes to their higher rates for stomach cancer. Meanwhile, rates of stomach cancer have plummeted in the developed world thanks to near-ubiquitous refrigeration, better sanitation, clean water supply, and food safety.

The flipside of public health's contribution to reducing stomach cancer is the phenomenal rise in lung cancer. Because of the evolution of base excision DNA repair and related gene networks, our lungs are now well equipped to handle that catastrophic toxic pollutant, oxygen, which invades our chests every second courtesy of Mother Nature's awful armies of oxygen-emitting plants. That breathing oxygen does not cause lung cancer was obvious until about two centuries ago. Before that, lung cancer was considered a rare disease, representing only about one in every one hundred fatal malignancies. Today it is the leading cause of cancer death, causing a little more than one in every four cancer-related deaths. In the early twentieth century, scientists suggested that several environmental airborne toxicants caused lung cancer's transition from rare to wicked: industrial smokestack emissions, automobiles, asphalting roads. Yet, the rates of lung cancer rose with similar velocities in countries with less industrialization, traffic, and asphalt paved roads.

Lung cancer's rise closely paralleled the introduction of cheap and mass produced cigarettes in the late nineteenth and first half of the twentieth centuries. Tobacco smoke mixes a potent blend of carcinogens with oxygen radicals. For reasons not presently well understood, only about one in seven long-term smokers develops lung cancer. It appears that genetic variation can have an important impact on protection from cancers as well as increasing risk.

Michael had grown up in a well-to-do Tri-State suburb, just outside New York City. His childhood was uneventful. Medically, both he and his family were generally, well, rather boring. This is the good kind of boring, the kind you don't appreciate until your life becomes cluttered with anxiety and emergency room and office visits and procedures.

The medical part of his life became far less boring, and not in a good way, one cold Chicago winter's morning. Then twenty-seven, he was awakened by wind howling off Lake Michigan and around his high-rise co-op. His bed sheets and pajama bottoms were bright with red blood from his rear. He went to the emergency room.

Overall, he had been going about his daily life and feeling fine, so he was genuinely shocked that he could have something seriously amiss. He felt well enough to hail a cab rather than call an ambulance for that trip to the ER. Michael is gay, and his immediate fear was that he had AIDS. Indeed, doctors at the ER first looked for an infectious cause, but he had been having no fevers, chills, pain, diarrhea, or other signs and symptoms of an infection.

To the surprise of everyone, Michael was found to have cancer in his colon. Only a few centimeters wide, the tumor had by misfortune opened a good-sized local gastrointestinal artery. This had caused the bleeding. It was an atypical and unpleasant way for the disease to present itself. However, it may have helped save Michael's life, since he was the first in his immediate family to have cancer. The entire clan was in shock. Both his parents and sister were cancer free. One of his grandmothers had previously been diagnosed with breast cancer, but that occurred when she was in her sixties. There was nothing in Michael's family history that announced he would soon gain expertise in human cancer genetics.

Because of his young age at the time of this diagnosis, Michael had genetic testing in Chicago. This work-up revealed that he carried a mutation in the gene called *MutYH*. This particular protein fixed a common problem. Oxygen or other chemicals damages the genome's Gs that are supposed to bind to Cs. When this happens, an A can be inserted accidentally instead of its proper mate the next time a cell divides, and the DNA is recopied. Consequently, cells from people with MYH-associated polyposis, or MAP, particularly cancer cells, have a precise molecular signature: instances of G:C are replaced by T:A. Tumors have fewer instances of G:C and more of T:A.

Identifying a MutYH mutation is an important actionable. These patients need frequent screening with scopes of the colon, the small intestine, and possibly the stomach because, with this disorder, growths called polyps can develop, with the risk that they will morph into cancer. These can be very few, or measure in the hundreds. If there are many polyps and these become unmanageable, the next step is surgery. However, the surgical approach employed is different from that usually undertaken for colon cancer, typically taking more normal tissue in order to reduce the chance of new tumors arising as polyps reappear.

Because this is a rare disease, it can be challenging to perform clinical studies of MutYH mutation carriers. While the evidence is presently limited, there may also be increased risks for cancer of the thyroid. The thyroid concentrates iodine, which is incorporated into thyroid hormone, and iodine is susceptible to being oxidized.

While MAP is rare, genetic lightning struck Michael not once but twice. First, lifetime risk of colon cancer in the United States is about 6 percent, with less than 10 percent of that occurring younger than thirty. Of colon cancer patients, only a few percent have a mutation in MutYH.

Second, he was mosaic. MAP is a recessive disease, meaning that both copies of the same gene must be mutated. This explains in part why Michael's family had no problems with colon cancer in earlier generations. His relatives had a working dominant version of the gene alongside the mutated recessive version. They were silent carriers of this recessive disease but otherwise largely no worse for wear.

Most people with the active disease inherit one recessive nonworking copy from their father, and the other from their mother.

However, Michael's particular MutYH mutation, inherited only from his father, had two mutated copies in about two-thirds of his body's cells. After conception, when his embryo consisted of a few cells, one of those cells inherited two copies of the particular strand of chromosome one with his specific MutYH mutation. Consequently, all daughter cells in the branches of the cell family tree in that fateful cell a few days after conception carried two copies of this recessive mutation.

Mosaicism, where cells with different genetic variants can simultaneously occur in the same individual, was first demonstrated to occur in animals in the 1930s, during research on fruit flies. In the 1950s, scientists found that it also occurs in people, with the discovery of what in those days were termed "hermaphrodites," but today are more precisely called "intersex" individuals, who have both male and female sex attributes caused by mosaicism of cells with either XX and XY chromosome pairs. Other examples of mosaicism include calico tomcats, which have mixtures of different coat colors, and likely the late rock star David Bowie, who had different-colored irises in his right and left eyes.

From this standpoint, we pretty much are all "mixed pictures," with various degrees of cell mosaicism with different rare mutations. And the genetic mixed picture that is each of us can change over time. The process of genetic mosaicism is thought to be an important aspect of the aging process and is the basis for cancer cells developing in the first place. Thus, while our genomes are *relatively* stable over our lifespans, the constant trickle of mosaic mutations makes the actual genomic stability of any given cell more of a theoretical concept rather than a reality. Recent studies of people over fifty found that about 2 percent have genetically detectable mosaicism in more than 5 percent of their cells.[4] This is likely just the tip of the iceberg. As technologies improve to detect genetic mosaicism, this number is likely to increase significantly.

Sizable studies including more than fifty thousand people have also established that the increased mosaicism that comes with the aging process elevates the risk of developing different cancers, most notably cancers of the blood, where risk is increased approximately tenfold as a natural consequence of the aging process.

Michael had the colon cancer surgically removed along with most of his remaining colon, since it was at risk for developing new cancers. After his recovery, he found the procedure had caused only a modest impact on his lifestyle.

He went back to work. Out of privacy concerns, Michael had his testing results performed anonymously and not included in his medical record. That he was mosaic and not all his cells carried the MutYH mutation means he was somewhat less likely than most others who carried the mutation to develop problems. Yet colon cancer at age twenty-seven reminded him that for each person, disease probabilities become binary events. You develop cancer, or you don't.

We discovered his mosaicism by analyzing his blood cells, but it can be difficult to extrapolate from that to determine what percentage of the cells lining his colon, stomach, bladder, and so on, carry this mutation. Scientifically and medically, it might be possible, but not very appealing from Michael's standpoint, to conduct invasive investigations to evaluate the cells in those other organs. Absent that, all we know is that there was at least one cell in his colon that had transformed into cancer. Therefore, from the empirical perspective, he accepted that he had to keep up with his cancer-prevention surveillance. Even after his surgery, he needed to have the remaining part of his colorectum scoped annually, and his stomach also examined about every five years. He also chose to have annual MRI of his abdomen and head to screen for other cancers, even though it wasn't very clear that this would benefit him.

Michael got back on his feet and didn't let his diagnosis slow him down. He struck out on his own and started his own investment fund, seeded by himself, family, and friends. He had spent years developing high-frequency algorithms designed to win that little more than half his derivative trades. This had put him deep into the top 1 percent of wealth. He said he used to have to tweak his trading strategies once or twice a year. But recently, as the space had become more crowded and competitive, he now had to revise his computer algorithms once a month or even more frequently. His fund prospered, albeit with some vertiginous draw-downs juxtaposed between mountain peaks, along with grey hairs along the way, he said. However, Michael, being a granular-detail kind of guy who had had conviction in the algorithms he wrote, kept his emotions in check and saw them through to the other side. I was soon to see just how he'd bring his granular, analytical sensibility to bear on his disease.

I met Michael in Manhattan at an unlisted restaurant off Madison Avenue in midtown. Not exactly a speak-easy, it was a white tablecloth, Continental cuisine establishment with tables set apart for private conversation and *fin de siècle* décor. It was in a brownstone townhouse with no signage outside. In fact, I still don't know the name or whether they have a listed phone number. Michael was tall, thin with a narrow face. He was blond with a slightly receding hairline and mustache. A bit formal, wearing an exquisite charcoal-grey-meets-cobalt-blue custom tailored suit, he spoke rapidly, articulately, with a precise, flat-affect Ivy League accent. His diagnosis of MAP syndrome and mosaicism had made him a "genetics junkie," a voracious reader on the topic. He had his genome sequenced at a concierge physician practice, and even requested a hard drive to keep his own copy, which also remained out of his medical record because of privacy concerns. He had spoken to a variety of other clinical geneticists and said he wanted to get my view on several issues. He had researched me heavily and even read some of my scholarly journal articles on colon-cancer genetics.

Michael had made a lot of money by trying to position himself to have probability working for him, rather than against him. He aspired to attempt the same with his genetic information. "I don't like playing poker blindfolded," he said.

He was fascinated by his personal genetic self-knowledge. Michael particularly enjoyed the irony that, despite his keen intellect, his Neanderthal DNA content was unusually high. Along with another 1 percent or so of male humanity, his Y chromosome apparently shared some ancestry with Genghis Khan. The sequence showed that he carries a gene variant that modestly impairs his ability to digest lactose, and indeed, he has mild bloating when he eats dairy products. Yet he readily accepted this was all cocktail-party talk.

It was the more serious medical variants he wanted to talk over with me. He was extremely concerned about wellness and disease prevention. He very much wanted his sequenced genome to be an integral part of this.

Because of his sexual orientation, he was at higher risk of developing hepatitis C. He carried a pharmacogenetic variant that confers better response to anti-viral drugs for this disease. He was disappointed that he didn't carry the strong HIV-protective allele in a gene called CCR5, but was glad to have this information rather than not know.

Additionally he had what a previous clinical genetics consult calculated to be a 20 percent higher risk of having a heart attack by age fifty. However, he disagreed with the method used to estimate this kind of risk because it involved multiple assumptions: he was also supposed to have increased risk for high blood pressure, but didn't have this so far. This made him suspicious of the statistical method used to determine such numbers. It was an exhilarating experience to have a patient who shared the same passion for the details of genetics that I did. I finished my lunch long before he did, and then sat there as Michael did most of the talking.

Michael understood that most genetic predispositions are complex traits. Genetic and environmental factors interact in such intricate ways that we generally don't have firm data as to what variants and exposures are doing what. He understood that without a strong family history for cardiovascular disease and no obvious mutations in the high-impact genes affecting heart disease it was very difficult to estimate his genetic risk from his precise, particular constellation of genetic variants, many of which had never been observed in the research literature before.

Regardless, Michael meticulously watched his diet and cholesterol levels. His approach was to estimate his absolute rather than relative risk of disease. And he was now confident that most of his disease risks were very low.

Yet, he couldn't make much sense out of what the physician who had ordered his genome test was telling him about many of his findings. He had done a great deal of his own research using PubMed, an online resource of scientific manuscripts sponsored by the US National Institutes of Health and accessible by anyone, with many of these articles available free of charge.[5]

Michael complained for quite a long time about the overall poor quality of too many genetic studies and how "we" (geneticists) weren't doing a very good job and should do better. He also complained about how the intelligence of most doctors wasn't very high compared to those in experimental finance and that so many were not properly quantitative, except when it came to ringing the cash register.

He was accustomed to the way huge volumes of known data could be crunched in experimental finance. He could come up with a trading strategy hypothesis, test it repeatedly on different Big Data sets to determine robustness, and do it quickly. Yet, he was frustrated by the lack of genetic Big Data, limiting his ability to do anything with most of his own genome despite his lust for this information. I opined that to fully take advantage of Big Data approaches, the global community would need to sequence the genomes of a million people and map all their electronic health records to the gene variants. I also suggested it would take a while to get people behind this idea and that it would require major seed money from a patron—unfortunately a role he didn't seem to be eager to take on himself.

However, the most important aspect of having his genome sequenced for Michael was trying to understand how this influenced his cancer risk. He was mosaic and was unsatisfied about why he had gotten colon cancer at such an early age, which could further imply a difficult-to-quantify higher risk of cancer elsewhere in his body. When about one out of five can live a whole life with a MAP syndrome mutation and still not get colon cancer, why did he get it so young? He had not been able to find in his genome a clear synergy as an explanation. There were too many unanswerable questions for his taste.

He then finally hit me with his major concern—the reason why, most of all, we were having this meeting. And it had to do with smoking: not cigarettes, but fish.

He wanted to understand whether his diet was affecting his risk. In particular, he went on and on about how much he loved eating smoked salmon. Since his childhood growing up in the New York area, he had loved eating smoked salmon, and he ate the best, every week. Michael had gone on treks sport fishing off Finland, Newfoundland, and Alaska to find the ultimate salmon. It was a passion, a near obsession, for him.

For people with intact DNA repair this was fine, but he had impaired DNA repair in his GI tract. Having scoured his genome for a cause of his cancer, he was now wondering if there was a gene-environment synergy. Had his mutation *plus* the volumes of smoked salmon he had eaten since childhood given him colon cancer? Diet appeared to have a significant influence on different DNA repair syndromes and even general colon cancer genetic risk as well. He had become familiar with the literature showing that the same families at high risk for stomach cancer a hundred years ago had stopped getting this cancer with the advent of refrigeration and reductions in nitrate salts in meat. He wanted to know: could he keep eating the smoked salmon he loved so much, or was this dangerous?

From Michael's perspective, the cost and inconvenience of a colonoscopy and MRI every year was perfectly acceptable. However, reflecting a quirky and all too common attitude among many of my own patients (reflecting, of course, people in general), his intellect and appetite were not in agreement. Even with a profound intellectual understanding of his entire genome and strong acquired background in MAP syndrome pathobiology, giving up something he enjoyed so much troubled him profoundly.

I told Michael that there are so many questions, and there is so much we don't know. We really have only scratched the surface. As crazy as it might sound to some readers, the smoked-food question was actually a very reasonable question. So many of these kinds of questions simply don't have clear answers yet, and there might never be enough patients to assemble meaningful data on Michael's specific question, even if it was of vital interest to him. He, or we, will not be able to approach genetic data the way that experimental finance can for a long time, possibly not ever.

So what could we do in the meantime? What to do when we have no useful data? We have to improvise and use our own personal risk-benefit equation for these known unknowns. That day at lunch with Michael, I began scribbling on the back of a napkin—very old school—listing all the known positives and negatives.

The cornerstone of excellent and effective clinical practice guidelines is *evidence-based medicine*. This term, abbreviated as EBM, was coined by physician, health-care analyst, and mathematician David M. Eddy, MD, PhD. In 1990 he defined EBM as "explicitly describing the available evidence that pertains to a policy and tying the policy to evidence" with the goal of treating individual patients.[6] Yet, there was at present very little evidence to help base Michael's care on most of his particular genetic makeup, MAP polyposis being the exception because it had a deep research literature to inform its management in the clinic. The relative ease of acquiring genetic data had run way ahead of the medical system's ability to acquire practical suggestions to inform Michael's care with the type of precision prevention medicine that he demanded. There would almost assuredly never be a clinical trial or multiple molecular epidemiology studies to answer this question, at present foremost in Michael's mind. Without that, in the meantime, we have to make our best guess. Eating smoked salmon might indeed be a serious problem for Michael, but we really have no way of knowing how much of an effect it could have. He might be able to eat tons of smoked food and never get cancer. He might get hit by a car or even by a plane crashing into a building long before any cancer even develops. Meanwhile, with 100 percent certainty he knows that he loves his smoked salmon.

Michael was frustrated that he had exhausted the current limits of genetic epistemology. In the end, his nostalgia for the tastes of his childhood triumphed over his towering intellect. He wound up deciding that, in the absence of definitive data, this quantitative risk-taker at the top of his probabilistic game would continue to eat his beloved smoked salmon. The subjective benefits outweighed the subjective risks.

I spoke with Michael about a year later. Two more fast growing colon polyps had been found in, and removed from, his residual colon tissue. He'd thought about it some more. He still loves his smoked salmon, but had changed his mind. He now eats it in smaller quantities, and saves it for special occasions. He's found that he now savors his smoked salmon more when he does eat it—less frequently, just in case.

FIVE

The Evil Twin

I was running late trying to get to an evening appointment, a dinner with colleagues to welcome a visiting oncologist and renowned colon cancer clinical trialist from the Mayo Clinic. The guest of honor was in town to give a lecture the next morning. As part of our hospitality, we were meeting at Felidia, an East Side midtown restaurant that had won Italian cuisine's equivalent of a Nobel Prize: hosting the Pope for dinner—in this case, His Holiness Pope Benedict XVI. By the time I got there, everyone else was already there at the table; I quickly heard about several interesting patients they had seen recently whom they were thinking of referring to me, the genetic-testing guy. One of these referrals particularly caught my attention.

Tim Reggers was a successful tech entrepreneur and software prodigy in his late forties. Originally from the New York area, he now spent his summers in the city and winters in Florida. He had pushed the needle in solar energy technology and material sciences, fusing these to build a better mousetrap for transforming light into electricity. After selling a couple of startups, he was now semi-retired. Tim and his wife had recently toured Northern Europe tech conferences and Scotland's golf courses for a month, on a journey that had been both productive and enjoyable, and picked up a new BMW along the way. But travel is always filled with the unexpected. New food, new experiences, new climate. Tim's arms, legs, and body had also starting feeling unusually itchy. He had a little bit of what seemed like the traveler's runs, but nothing out of the ordinary. Tim was in no distress and in general felt pretty good, even if a little less hungry than usual.

Returning from Europe to New York, he transferred between flights at the Amsterdam International Airport. Tim's wife noticed his skin looked unusually dark. While Northern Europe is a leader in solar power technology, it has never been known as a mecca for tanning. In fact, the weather during their trip had been rather overcast and rainy. It was there, at the airport, that his wife had also noticed the whites of his eyes were, well, not so white. They were yellow. In the airport, he went to the medical clinic. There, a calm urgent-care doctor told him that he had hepatitis, which had caused his jaundice.

Jaundice occurs when the liver isn't functioning properly. The liver gets rid of old red blood cells by making them into bile, which is flushed down into the intestine. This was the most likely explanation of his having "the runs," the yellow whites of his eyes, and bronzing skin. He didn't seem to be in any acute distress except for a slightly diminished appetite. He was told to follow up with his doctors in New York after he arrived at JFK Airport the next morning. All of this seemed quite matter-of-fact, and not particularly urgent, and to Tim it seemed a lower priority than visiting his elderly parents after he returned.

When Tim arrived back in New York, he saw his internist, who confirmed that his liver enzymes were elevated. A constellation of tests suggested that the liver was actually working properly, but that it was not able to properly dump its bronze-hued bile into the small intestine. This suggested he might have small gallstones blocking the outlets in a structure called the *biliary tree*, a condition that usually, but not always, leads to pain in the right upper quadrant of the abdomen. Yet on his abdominal CT scan, gallstones were nowhere to be seen. Nor did the scan show anything else amiss. No other masses, no enlarged lymph nodes. The branches of his biliary tree were somewhat wider than expected, but this finding is very nonspecific. In summary, there was nothing to raise his doctor's eyebrows.

Tim's mother had developed cancer of the stomach when she was in her forties. It had fortunately caused symptoms and, since the alarm bells rang early, it had been caught and successfully treated. But the coincidence of a close relative with gastrointestinal cancer in what was a small nuclear family gave his family physician pause.

Tim's family history, combined with other uncertainties about the diagnosis, prompted the doctor to call for a procedure called an *endoscopic ultrasound.* Endoscopic ultrasound, also known as EUS, involves using a thin, flexible scope that goes down from the mouth into the stomach and then into the small branches of the bile ducts connecting the liver and gall bladder to the small intestine. This arbor-like network of branches dumps bile into the intestine to help with digestion. EUS uses a tiny ultrasound device at its tip to hunt for anomalies such as cancer cells. The ultrasound showed that there was indeed a mushroom-like bulge in the biliary tree near the gall bladder. The tissue from a biopsy of that bulge revealed agitated-looking cancer cells.

Scanning didn't show any signs of anything abnormal in any other tissues, so Tim was scheduled for a *Whipple procedure*, named after the American surgeon who, in the 1930s, improved upon an already-complicated surgical procedure first described in 1898.

More than a hundred years later, we are still using this surgical procedure. The modified Whipple procedure consists of removing the lower half of the stomach, gallbladder, and its ducts into the biliary tree, along with about two-thirds of the pancreas, the upper several feet of the small intestine and lymph nodes (where tumor cells draining from these tissues can be filtered out and killed by the immune system). Next, the reconstruction process begins. The remaining upper part of the stomach, the tail of the pancreas, and the liver are attached to the rest of the small intestine.

It is one of the most demanding of major surgeries and can last hours and hours. Going under "the Whip" demands great stamina on the part of the patient, as well as the operating team. About one out of every twenty to fifty patients (depending on the particular surgeon and hospital) who have this operation does not survive, even today. There can be loss of six or more units of blood, and the removal of most of the pancreas can complicate maintaining stable blood sugar level.

The goal of the surgery is to cause a complete cure. Because Whipple procedures are so complicated and risky, they are only performed with the goal of removing every last cancer cell. The trauma and risks are not considered worth taking if the cancer is likely to return anyway.

The procedure can sometimes leave the patient diabetic if too much of the pancreas—whose islet beta cells make insulin to control blood sugar—is removed. But if all goes well, the digestive system, and the patient, can begin a long healing process, which can eventually be successful with a great deal of patience and support from loved ones and an engaged rehabilitation staff.

As a medical student, I had participated in several Whipple surgeries. After the opening of the abdominal cavity, the job of the medical students who rotated on and off the surgery service was to hold for several hours the surgical retractor that helped to keep the large incision open and all the different organs visible and accessible to the surgeon. We were advised to avoid locking our knees and standing too straight, because after several hours this can cause blood to pool in the lower legs and has induced rotation students to faint in the operating room.

Tim's modified Whipple surgery lasted more than four hours. In the end, everything went according to plan. Most of the pancreas, the first part of the small intestine that attaches to the stomach, and the biliary tree were excised *en bloc* (that is, in one piece) by the Whipple procedure.

The team in the operating room had succeeded, but there remained much more work to do, while the excised tissue was passed on for visual examination and other tests by the surgical pathologists. While surgical pathologists don't get the attention or glory of star surgeons, an observant and smart pathologist behind the curtain can mean the difference between life and death for the patient.

The art and science of surgical pathology is at its essence one of pattern recognition. Intuitively, surgical pathologists develop a feel for different diagnoses after looking at thousands upon thousands of slides, each having thinly cut sections of tissue. It is somewhat like an apprenticeship of the microscope. Similar to an expert art critic, each individual pathologist's point of view and ability to pick up patterns of diagnosis have been molded by mentors. The mentors, in turn, were taught by their own mentors. Thus, each hospital can have its own institutional culture of surgical pathology diagnosis, with little use of universal criteria or diagnostic "gold standards" from one hospital to the next.

Subjectivity as well as objectivity is weightily represented on the scales. For many diagnoses, particularly for rarer diseases for which there is less of a collective experience base, there is more agreement between pathologists from the same hospital in coming to a diagnosis than there would be from pathologists from different institutions. In addition to the effects of mentor-driven institutional cultures, there can be a sort of group bias driven by a "big dog" effect, in which a group of people in a room looking at something and trying to agree on a conclusion defer either to the most "experienced," alpha-dog, senior physicians, or to the loudest, most articulate, most opinionated physicians.[1]

When the surgical pathologists examined under magnification the part of the biliary tract called the *Ampulla of Vater*, where the bulge was visible in the endoscopic ultrasound, they saw a lump jutting up above the bile duct. The growth was about the length of a toenail (about 4 cm), looking like broccoli flower, and below another centimeter-sized clump of abnormal cells biting into the major bile duct. There also was some good news: no signs of cancer cells that had metastasized.

The pathologists processed tissue from the lump using essentially the same methods as had been used for more than a century, staining the cells with chemicals originally used for dyeing women's clothing and staring at this stained sample under a microscope brightly lit from below. They classified these cancer cells as "high grade." Their nuclei looked like large, dilated, and angry eye pupils. They were bulging with disorganized, irregular looking chromatin, the proteins that pack the DNA together like spools of yarn. Like zombies, they had been transformed into something unholy, a consequence of a genome deranged.

Because tumor cells make chemicals that distinguish them from normal cells as they breathe oxygen, take in food, and divide to make daughter cells, the coloration of the cells was also abnormal. The tumor cells were darker hued. Their grainy cytoplasm was smoky vermillion. The rest of the tissue was processed with an acid called formalin and wax to keep the cells and surrounding tissue fixed in place, then sliced to only about the thickness of a human hair and placed on special sticky slides for more detailed examination by a surgical pathologist. The visual examinations and other tests for proteins confirmed that these cells were consistent with cancer arising from the cells that line the biliary tract. Another type of lab test, using antibodies to look at the presence and absence of different tumor proteins, helped further confirm that this appeared to be a biliary tract cancer and not a metastasized "tourist" from a cancer that had arisen at another location. This is a difficult type of cancer to treat. Bile is produced by the liver to help digest food, and the cells lining its ducts are especially adapted to survive in a hard, acid environment. Like zombies, cancers that survive in the biliary tract tend to be difficult to kill off.

The next question: *exactly* what type of cancer was this? The answer was critical because that would dictate specifically which anti-tumor drugs Tim would be given.

Once upon a time in the twentieth century, the diagnosis would simply have been "cancer of the bile duct." It is often said that the history of medical research is a tug of war between two camps of doctors: lumpers and splitters. To understand different diseases, doctors have to organize them, whether by signs/symptoms, etiology, or underlying mechanism. Lumpers look for commonalities between different diseases to classify them. Splitters stare at diseases and try to see how several diseases that superficially look similar in fact have distinct etiologies. Overall, the splitters have pulled way ahead during the past few hundred years.

In my practice when taking a family history I often hear that someone's grandmother died "from cancer of the abdomen," with not much more detail than that. Then, that the patient's father died from cancer of the kidney or colon. Those lumped descriptions reflect the state of medicine at the time with less precise knowledge of the different types of neoplasms that can arise in the same organ. Today, the common practice of naming cancers by the organ or origin (lung cancer, colon cancer, breast cancer, and so on) remains the norm. However, the fairly recent recognition that cancer is primarily a disease of DNA has begun to change the classification of cancers, organizing them more specifically according to the mutations carried by the tumor cells.

A major clue to the specific type of cancer came from what the pathologists also found: Tim's cancer had been invaded by large numbers of the distinctive-looking immune cells called lymphocytes, small round cells with little cytoplasm, since almost all of the cell is taken up by the DNA-containing nucleus and nucleolus. They are part of what is called the *adaptive immune system*. The adaptive immune system evolutionarily is thought to have originated in ancient jawed fish approximately five hundred million years ago to fight off infections, as well as possibly cancer, although similar cells are seen in simpler animals such as the radially symmetric sea urchin. Sometimes you have to fight fire with fire, and the adaptive immune system uses mutations for the good: to fight our foes and repel foreign invaders, including bacteria, viruses, but also cancer cells.

It is Darwin's war, as each side deploys the weapon of mutation and evolution to conquer the other. Elite special forces of tens of millions of lymphocytes called T and B cells have, on their cell surface membranes, proteins that are altered by different combinations of mutations to generate what amount to three dimensional "locks" with almost every conceivable shape. This process is called *hyper-mutation* because it allows a small number of genes making these cell surface membrane proteins to generate a huge and diverse army with literally billions of different combinations.

In the most sophisticated version of precision medicine ever seen, the body's T and B cells circulate over every corner of the body, looking for a fit with foreign "keys"—in the form of foreign invaders. Almost none of these cells ever find an uninvited infectious invader with a matching shape. However, with the help of command-control helpers called *antigen-presenting cells* that help the T and B cells find their key, when an extremely rare and lonely lymphocytes does find a fit with a novel shape not normally present in the body, the antigen-presenting cell promotes the lymphocyte, which initiates a process called *clonal expansion*. In this process, a handful of cells divide rapidly thousands of times to produce an army of "armed and ready" lymphocytes to protect the body. This adaptive immunity army recruits additional immune cells to participate in the attack.

When these T and B cells congregate around, bind to, and react against foreign shaped proteins and other molecules on cancer cells, they are referred to as *tumor infiltrating lymphocytes*. Tumor infiltrating lymphocytes are a good sign in two ways. First, they indicate that the host patient's immune system can function well and react against the foreign tumor shapes (as many cancers simultaneously fight back by suppressing the immune system generally as a whole using a multitude of sneaky tricks). Second, they offer a signal that this particular tumor's cells have a shape on their surfaces that the tumor infiltrating lymphocytes can recognize and actively engage in fighting. Overall, the presence of tumor infiltrating lymphocytes is a good sign. Their multitude increased the odds Tim would do well to win this struggle.

The number of Tim's tumor infiltrating lymphocytes in his cancer was unusually high. The way that they crowded against each other, pushing and shoving to get at the cancer cells, pressing up against their foes in hand-to-hand combat, was a pattern the surgical pathologist had seen before, one often seen in Lynch syndrome colon cancers. As discussed earlier, because they are caused by mutations in DNA repair genes, Lynch syndrome tumors have mutation rates at least ten times higher than normal, so they produce many more foreign shapes that the adaptive immune system can recognize. Consequently, they are known typically to have very high numbers of *tumor infiltrating lymphocytes*, or TILs. While the biliary tract cancers in Lynch syndrome are quite rare, occurring in at most a few percent of patients, the pathologist's finding, combined with a family history of gastrointestinal cancers, prompted genetic testing of the patient.

His oncologist referred Tim to clinical genetics, and I saw Tim and his wife in clinic several months after his Whipple procedure. He did not have any sebaceous adenomas (which are usually benign skin growths associated with Lynch syndrome that can on occasion tip us off regarding the DNA diagnosis). Tim looked energetic and in good spirits, and had been gaining weight. He carried the look of determination of someone who would do whatever it takes to get well. The stare in his eyes was angry: he was mad at his cancer, and it was obvious that he wanted to pound it into the ground and be done with it. That kind of enthusiasm is infectious to a doctor, and the patient becomes teammate. We performed testing of blood DNA, and Tim was found to carry a mutation in the MSH6 gene, one of the less common causes of Lynch syndrome. His mutation wasn't in the part of the gene that encodes the protein. Instead, it was in adjacent DNA that affects how the RNA is spliced together.

The central dogma of genetics is that DNA is made into RNA, which is then translated by ribosomes into proteins. For many genes, including Tim's MSH6, the RNA is produced in pieces that are then sutured together to encode the full protein. In this case, the mutation caused one piece of RNA to be spliced out, which caused the rest of the protein to fall apart.

Normally, this is not a problem, since as with most genes there is a backup copy that kicks in and picks up the slack. On a microscopic level, this is similar to the reason we have two eyes: if the right is injured, we have the left so that we can still see. However, somewhere along the way, in some errant, originally normal-appearing cell deep in the arbor of his ampulla, his second MSH6 gene somehow got broken. Consequently, MSH6 could not perform its role as a guardian of the genome in that single, fateful cell that changed his life. As that single cell divided repeatedly, it quickly reproduced its sequence with a multitude of errors that ultimately led to tumorigenesis: the birth of a cancer cell.

The many errors explained why there were so many tumor-infiltrating lymphocytes. This particular tumor was making proteins that had lots of typos because it had lost the MSH6 spell-checker protein. The TILs were reacting against these new shapes decorating the cells' surface that Tim's immune system had never seen before and attacking them like a cell infected with a virus.

Recovering from a Whipple procedure is not easy, but Tim had slowly and surely improved. Enough of his pancreas remained after the surgery that his ability to make insulin and control his blood sugar was fine, so that he didn't have any problems with diabetes. Eating well was a little harder, as his gut was more sensitive to what he ate. Since he was a little more susceptible to losing the digestive enzymes of the pancreas, Tim took pills containing the pancreatic digestive enzymes four times a day, and he had gained about fifteen pounds since the surgery.

Then he had the dreaded rite of passage for all cancer patients: the day for imaging tests to see if the cancer had returned. Tim and his wife exhaled mightily as they found out that the CT scan of his abdomen revealed no sign that any cancer had returned.

Tim continued to do well and was gaining more strength back and hoping to put all this in the past and move on with his life. Because his

Lynch syndrome had been identified, he had screening colonoscopy and stomach endoscopy and could to plan for more screenings in the future to keep any other Lynch syndrome problems at bay.

It was late summer, when New York was lazy, tired, and (relatively) empty. Uber and Lyft rides from "Bedpan Alley" (the neighborhood on Manhattan's Upper East Side filled with hospitals) were off peak, cheap, and plentiful, for a change. On a rainy and otherwise unremarkable day six months later, Tim returned for more tests. The results were not as we had hoped. He had three new unidentified bright but irregular and ragged new dots not far from where his Whipple had been, on his liver. Another procedure confirmed what none of us wanted to hear: Tim's cancer had reemerged out of the dark. While the liver's bile flows down into the biliary tree, some of the cancer cells apparently had climbed upstream into the liver, or perhaps crawled around the surrounding squishy connective tissue and gone into the liver through the back door. On the brighter side, there were no signs of tumor that had spread beyond the liver—at least one sign of hope.

Tim's life then entered the next stage, the complex world of chemotherapy. The standard drug combinations for cancer of the biliary tract were not what he wanted, on average extending life for only a few months and truthfully usually not anticipating a cure. He was feeling well and went for a powerful newer combination with the difficult to pronounce acronym of FOLFIRINOX. The ten letters in this acronym indicate that this treatment incorporates many drugs (in this case four) and packs a wallop. My patients taking this type of chemo mélange have said that the fatigue and soreness it can induce are "like getting kicked in the chest by a large mule." It includes, among others, oxaliplatin (a DNA-damaging agent that originated from the observation that bacteria were killed when exposed to the metal platinum), irinotecan (a drug that permeates pretty much every cell in the body and prevents the unwinding of the spools of DNA when cells replicate), and 5-fluorouracil (which pretends to be the nucleotide thymidine and is accepted into the genome but causes DNA to become fragile and shatter from within). These drugs would kill the most rapidly dividing cells in his body, the opening salvo in the war against Tim's rare cancer. This is one of the less commonly used frontline regimens for chemo-naïve (that is, never before treated with chemotherapy) tumors, in part because it was difficult to tolerate the side effects, which can be significant.

However, the team's overall assessment was that the potential benefit of a tantalizing, rare cure for bile duct cancer outweighed the pain. While no one likes being kicked by a mule, these elixirs held out the small but perceptible chance for a cure. It was a kick with a good purpose.

Tim received four cycles of FOLFIRINOX injections. He took it all in and fought back with optimism, determination, and a competitive drive to beat his cancer. Yes, he was fatigued, sore, and bruised from his fight, but the liver tumors all had disappeared, except for two. While X-ray fluoroscopy imaged his abdomen, a catheter was positioned adjacent to the dreaded spots, and high-frequency radio waves were used to ablate these remaining growths.

Three months later, he was feeling well, just suffering a little muscle cramping. His appetite had returned, and he was gaining weight. Tim's next MRI in the summertime showed no sign of anything abnormal.

Then in the fall, his next imaging showed that cancer cells had grown back at the sites where the two spots had been zapped. Tim went through another round of radiofrequency ablation. Then he had another round of chemotherapy, this time with fewer drugs that were easier to tolerate. Still, it was clear something else had to be done. The malignant terrorists were fighting back, constantly moving, adapting, and changing the rules of the game in real time.

What often feels like a *one-size-shoe* approach is the legacy model of oncology for treating each type of cancer. For common cancers, the cornerstone of choosing which drugs to give patients is evidence-based medicine, built on large-scale trials. The precise chemotherapy cocktails used are fundamentally derived from rational treatment strategies based on these heroic efforts. These trials (of course, organized around the organ sites from which the cancers arose) can involve many thousands of men and women and use statistical tests and rigorous and meticulous measurements, even to the point of detecting differences of only a few days of life between average patients among different treatments. These studies can often provide definitive data on how doctors should best manage patients, given the limitations of the drugs that were available during the study.

This approach makes a great deal of sense as foundation of national health-care policy and for determining which drugs to pay for. Indeed, well-performed large-scale clinical trials and evidence-based medicine were some of the greatest achievements of the twentieth century, propelling progressively expanding life expectancies. However, such trials are enormously expensive, sometimes running to more than ten million dollars per definitive clinical trial. They also take a long time. Consequently, there are many questions that we just don't have enough solid, definitive data to answer.

That's part of the reason, from the viewpoints of the patient and treating physician in the battlefield trenches, the one-size-all approach doesn't always look as appealing. For each individual with a new diagnosis of cancer, the outcome is not a statistical average of hundreds or thousands of patients but a singular event. To come up with the best cancer treatment plan, that singular individual doesn't want the average solution for the virtual average person on the standard institutional home screen, the turbid brown from all thousands of subtly shaded colors mixed, but rather a solution precisely personalized for him or her. The twenty-first century demands it. Our patients demand it.

Concerned by the odds of winning by playing by the rules, Tim and his family wanted to do whatever it took to give Tim his life back, to use technology to build a time machine to take his life back to the way it was before THE diagnosis. Tim encountered a bewildering array of potential options. Should he stay close to home, maximize his quality of life and time with his family and friends, and accept his fate? Should he pursue experimental drugs and commute back and forth from Boston? Should he seek out new RNA-based therapies attempted in Texas that readily cured mice with cancer but untested in people? Or similar gene-based treatments in California? To become an expert in biliary tract cancer quickly and determine the optimal plan of attack, Tim sought and weighed different opinions about what to do next. He was evaluated and heard potential options in parallel at the Dana-Farber Cancer Institute, in Boston, and both Memorial Hospital and New York Presbyterian Hospital, in New York, practicing shuttle medicine from LaGuardia and Logan airports. This way, multiple options could be considered and debated. Detailed discussions weighing risks and benefits would help insure that a navigation plan on the frontiers of knowledge to eliminate Tim's micrometer-sized nemesis was well vetted, on course, and on point.

Then, in order better to triangulate between doctors and records at different institutions, and even different continents, Tim and his family hired a private concierge health-advisory service based in New York, PinnacleCare. (One European patient's family described the strange, complicated terrain of the American health-care landscape as "extraterrestrial.") To help navigate the complex terrain, the advisory service acts as a kind of medical version of a personal financial advisor, overseeing health rather than wealth. The service vets potential treating physicians, gives independent medical opinions, gets patients into hard-to-see specialists, tries to mitigate risky decisions, and does whatever else is needed—for a fee, of course.

A former veteran critical-care nurse was assigned to act as Tim's patient navigator. He accompanied Tim and his family to clinic visits, acted as a translator from oncology and genetics to plain English, and relayed data back to the advisory doctors. They in turn conferred on the case and advised Tim what to do next.

A true believer in the transforming power of modern technologies, Tim gravitated to identifying the precise genetic mutations in his tumor to help treat his cancer. Since we now know that cancer is fundamentally a disease of damaged DNA, this knowledge is changing the paradigm that physicians and scientists use to approach the diagnosis and treatment of cancer. We also now know that at the genomic level there is a frightening amount of heterogeneity among human cancers, sometimes seemingly as much as among the people who inhabit New York City. In the context of the thousands and thousands of genetic mutations carried by each malignancy, every person's cancer, whether from the ampulla, breast, or anywhere else, has many unique features. Remarkably, the medical world may never have seen many of these features in any other patient.

The astonishing genetic diversity of these tumors is not well captured by clinical-trial statistical averages. How many patients are outliers in how they respond to a therapy, and how far out there are they anyway? Each man and woman wants to know: if the common potions aren't a cure-all for what ails me, are there other mixtures sitting somewhere on a shelf that might vanquish my cancer? Perhaps that's a chemotherapy agent not commonly used for my particular organ's cancer but used successfully for some other type of malignancy, or even another disease entirely. Patients like Tim are looking for an arrow dipped in the exact venom to kill precisely his unique cancer.

From many sources, including radio and Internet ads, friends, family their hired health advisory service, Tim and his family were hearing a lot about *precision medicine*, a new strategy to attack tumors. This was a different kind of animal than conventional cytotoxic chemotherapy care, more cyborg than sinew. Precision medicine aspires to data-mine the idiosyncratic, sometimes unique particular genetic mutation fingerprints of each cancer. It compiles massive amounts of tumor genetic data to probe for the cancer's often-rare Achilles' heels in the hopes of finding the best line of attack therapy. Using algorithms and powerful computers similar to those used in analyzing drone video images in the search for Osama bin Laden, the basic idea is to scan not only clinical-trial middles, means, and medians, which doctors can readily see from the back of the auditorium in the center of enormous, bright high-resolution screens, but the dimly lit statistical tails on the corners of the periphery, where tantalizing possible cures reside.

Tim signed his patient consent form, and our hospital mailed off samples of his biliary cancer. Tim's oncologist and the hospital sent slides with micrometer-thin slices of the cancer for analysis to a new startup molecular diagnostic company in Massachusetts. The cost was about six thousand dollars. His insurer balked. However, in a quirk of our free-market medical system, the company ran Tim's test anyway, even though no one paid for it. (There is a long history of biotechnology companies that lose money in the short run but assume dominance over what is likely a fast-growing niche, in this case tumor genetic diagnostic tests, and play the long-term game.) Genetic testing without billing can be a great benefit for patients who are early adopters of still-emerging state-of-the-art technologies. The dull, semi-translucent sticky tumor tissue on the glass was then micro-dissected away from the non-cancerous tissue under high magnification by a surgical steel scalpel and scraped into a plastic tube. Phenols, alcohols, salts, and other compounds were added to purify the DNA away from the cell jetsam and flotsam. The almost invisible dot of DNA at the tube bottom was then amplified and separated by sliding the DNA fragments through the pores of a bed of agarose gel, harvested from the Pacific Ocean's kelp "forests" off California. Then protein enzymes, including one from a bacterium originating from the thermal geysers of Yellowstone Park, Wyoming, were added to prep the DNA for sequencing. After a little more pampering, trillions of short DNA reads from Tim's tumor were grown in a data farm hothouse of tiny silica beads and iteratively photographed thousands of time to harvest each bead's precise sequence bounty.

Tim's cancer biopsy was massively parallel-sequenced for mutations in just over two hundred genes. Even though this sample amounts to only about 1 percent of human genes, the two hundred–plus we targeted had something critically important in common: they are the known actionable genes. As today's actionables, they form the focus of the new way of thinking that has begun to classify cancers more precisely than does organ-centric criteria.

If an actionable gene is identified, it can mean that we can use an existing medicine to block the cell-signaling pathway that would otherwise trigger the cancer's proliferation. For example, drugs that inhibit a mutant *BRAF* gene driving cancer cell proliferation are not commonly used in bile duct cancers. However, if this analysis showed that Tim indeed had that mutation, he might the lucky exception for whom the targeted BRAF drugs might work. It could mean that his disease had a chance to be kept under control by an easy-to-tolerate pill, and perhaps he could lose the plastic chemotherapy port in his right chest and be spared another kick by that supersized mule. A mutation in another called *AKT1* could mean having to go to a hospital in Germany or California to enroll in a clinical trial and receiving a newly synthesized small molecule therapy that targets this gene (or perhaps just the bitter pill of a placebo). And if the cards drawn showed a combination of druggable mutations, well, that could be like having a royal flush and raking in the jackpot.

Why not sequence all of Tim's roughly twenty thousand genes for about ten to fifteen thousand dollars? After all, that's less than the cost of many precision-medicine targeted cancer drugs or a week in the hospital. Doctors love data. However, the scales tipped on the side of "less is more" for several reasons.

Remember that the cancer actionables are the Holy Grail of the precision medicine quest. The only real issue of concern to Tim and his family was this: precisely what drugs should be used to kill the tumor? If we had chosen to sequence all or most of the tumor genome, we would have wound up with vast terabytes of information that would have been of little or no help in choosing a better therapy for him. This is a common contemporary theme in precision medicine. Our ability to acquire genetic information still *far* outstrips our ability to understand the significance of the findings in order to better treat a patient.

In addition, DNA sequencing cognoscenti know well that there are important types of genomic mutations that can be missed by the, at present, state-of-the-art DNA sequencing technologies. To make genes, all the DNA sequences are stitched together by computer programs in a manner conceptually similar to the way that pieces of a jigsaw puzzle are fitted together to make a complete picture. However, most of our genes have two copies. So, imagine you have a jigsaw puzzle with two copies of each piece, and just to make it more fun, a bunch of random pieces that don't fit in the picture at all come along for the ride (because there are always poor-quality DNA sequences that get jumbled with the high-quality ones). If at the end after you have all the pieces put together and one copy of a gene has a piece left over, it is usually obvious where it is supposed to go. However, if a piece has sequence changes far from normal, it can be tricky to figure out where it is supposed to go in the first place, and it just gets thrown into the junk piece pile. Thus, this type of mutation with many changes can in reality be present in the tumor, but unfortunately missed by the analysis.

Recall the surgical pathologists peering at slides of Tim's tumor through their microscopes. Just at the pathologists play a critical role in evaluating disease tissue, computational biology specialists called molecular geneticists play a critical role in evaluating the trove of information that pours from the DNA-sequencing apparatuses and the computers linked up with them. Unfortunately, on this frontier of science there is a wide variety of competency among clinical molecular geneticists. It can help greatly to have a mechanic who really knows what to look for under the hood.

For Tim, what we did *not* need were molecular geneticists who were the equivalents of jacks of all trades, but masters of none. We wanted these computational biology specialists looking at the sequences to be experts used to recognizing the subtleties of picking up mutations in the same genes that interested us. Although there's been no clinical trial so far to back up this strong hunch, my own experiential hypothesis is that the quality of the data regarding the actionables will be better if we choose the labs with the best of these experts and allow them to focus on fewer genes.

For example, another clinician approached me about a diagnostic dilemma. In a little over a month, he had seen three unrelated patients with colon cancer whose tumors had evidence for mutations in Lynch syndrome genes, but their normal DNA from blood (that is, non-tumor DNA) had each tested negative for a multi-gene hereditary cancer risk panel. Based on my experience, the tumor findings, and their personal and family histories, I anticipated that at least one of them would have had a mutation in a Lynch syndrome gene revealed. Suspicious about these three mutation detection swings-and-misses, I contacted the director of the molecular diagnostic laboratory that performed the gene panel sequencing and asked him to re-examine these patients' primary data. Greater scrutiny of their gene panel sequencing identified on second review a somewhat more subtle type of mutation to detect (one that deletes one of two copies of several parts of the gene that encode protein in one of the genes, leaving only normal DNA sequence) in one of the patients for the MLH1 gene and confirmed it with additional experiments. In Lynch syndrome, mutations like this called "deletions/duplications" cause as many as one in four or five cases of the disease. Thus, in busy molecular genetics testing laboratories, I sometimes wonder whether ordering more focused panels leads to better quality.

Tim's genetic quest for actionable targets turned out to be a success. A search of FDA approved drugs in the United States revealed that two mutated genes in his tumor, called ERBB2 and EGFR, had readily available drugs targeted against them, some others already approved by the FDA, and more in the pipeline as clinical trials.

Trastuzumab is one of the first targeted chemotherapy drugs, an antibody therapy that binds to and blocks signals from the protein made by the ERBB2 gene. Originally called *neu*, ERBB2 was discovered in 1979 by MIT scientists as being important for the growth of multiple types of cancer. However, because a higher percentage of breast cancers express ERBB2 compared to the other types of cancer, and the traditional oncology paradigm is of organ-centric cancer, trials for breast cancer were given the top priority.

Ironically, despite its obvious importance as a drug target, the original laboratory preclinical studies of ERBB2 in the 1980s showing it to be a likely important druggable target were not replicated in studies by several competing laboratories. This delayed development of drugs targeted against ERBB2. This inconsistency was later attributed to "use of contaminated chemicals, faulty techniques, and idiotic mistakes by the laboratories conducting the experiments."[2] Anti-ERBB2 drug development languished for several more years, until the mother of a vice president at the biotechnology company Genentech was diagnosed with breast cancer, and the company then reinvigorated its research program on ERBB2. The original work was replicated and validated to prolong survival of breast cancer patients, the gold standard in oncology. Now there are three drugs focused on ERBB2 inhibition and several others in the pipeline. We also now know that, as originally predicted, ERBB2 is highly expressed on a number of different types of tumors in addition to breast cancer.

When Tim had his next round of chemotherapy, in addition to his regular drugs, trastuzumab was added. Given the evidence at hand, his insurance paid for it as well, even though a randomized trial for using this drug to treat biliary cancer had never been done. As predicted, Tim's tumors in his liver shrunk and he felt better. His tumors did not completely disappear, but he was feeling well and gratified that there were now options where none had been before.

Limiting the number of genes we test for can have an entirely different sort of advantage: we might be able to avoid the incidentaloma syndrome. The incidentaloma syndrome refers to the problems that arise from finding something we weren't initially looking for and that all too often becomes nothing more than a distraction. Sometimes such a distraction can have serious unintended consequences.

This is a common problem in medicine and not specific to genomic tests. For example, a healthy guy hurts his shoulder playing tennis. His doctor orders an MRI of his right humerus, but the patient wants a full-body scan, "just to be safe." Why not, if the insurance company will pay for it? *Caveat emptor.* Well, the scan finds a smudge in the right lung. This leads to an interventional radiologist poking a needle into the right lung to biopsy the ditzel, which turns out to be a clinically unimportant nodule-nothing. Good news! But in the process of guiding the needle, the doctor accidentally punctures the lung, collapsing it, and gets the poor guy admitted into the hospital with a chest tube. After he recovers from the collapsed lung, he has to wait a while to get his game back on the court. The benign lung nodule would never have given the guy a problem if he hadn't insisted on getting the full-body MRI, which was happily provided by a fee-for-service-driven medical system that thrives on these impulses.

Incidentalomas are particularly problematic when complex tests, like genome sequencing, are run. In fact, sequencing a human genome is essentially trying to do three billion tests simultaneously. Yet, the incidentaloma syndrome can come into play even with tumor gene panels involving a couple hundred, rather than thousands, of genes.

Another patient of mine, Veronica, had a similar tumor genetic sequencing test for ovarian cancer, which she'd been afflicted with at an early age. In the genes on her tumor panel, there was only one that was called a mutation. This single letter change was in the TP53 gene. This is one of the most famous genes in the world. There are thousands of scientific papers written on TP53, which has been called the "guardian of the genome." Entire scientific careers have been dedicated to understanding its mechanisms. In fact, because it is such a heavily studied gene, TP53 has its own individual mutation database, a rare resource not available for most genes.

The TP53 protein's role is to ward off the development of cancer. It first helps assess whether the quality of cell DNA is good or bad before a cell divides. It then helps prevent cells with many mutations from dividing and thereby reduces the risk of a transformed cancer cell arising.

But, paradoxically, when TP53 itself is mutated, cells with damaged DNA instead can grow in an uncontrolled, unchecked manner, forming tumors. So in its mutated form, this otherwise critically important anti-cancer gene is also implicated in a wide variety of cancers.

A search for clinical trials relevant to Veronica's tumor mutations came up empty.[3] While TP53 mutations are associated with cancer chemoresistance to many standard anti-cancer drugs, there was still a chance that her tumor would respond to the chemotherapy even with a TP53 mutation.

There was no clear path forward. Unfortunately, despite the tremendous wealth of knowledge that we have about TP53's multiple roles in different mechanisms of tumor suppression, finding it mutated is not presently very helpful in terms of pointing to the best treatment option. On the frontier of knowledge, oncologists have to go with their gut instincts about what is best for each patient. This is still personalized oncology as much as precision oncology. I am confident that at some point we will get around this problem and make effective therapies to restore TP53 function. Unfortunately, that day has yet to come.

Finding a TP53 mutation in Veronica's tumor also opened up an incidentaloma box. The test had sequenced her tumor and discovered a *bona fide* TP53 deleterious mutation. Yet we couldn't tell whether the mutation had only recently arisen as the tumor grew and changed (called a *somatic* mutation), or whether Veronica carried this mutation in all the normal cells in her body (that would be called *constitutional* mutation).

People who carry a TP53 mutation in all the normal cells in their body have a genetic disease called the *Li Fraumeni syndrome*, named after Fred Li and Joe Fraumeni Jr., the American physicians who first described it. The Li Fraumeni syndrome can be passed on from generation to generation, although about one out of five or ten people with the syndrome have a *de novo* mutation and are the first affected member of their family. Members of Li Fraumeni syndrome families have an approximately twenty-five-fold increased risk of developing a malignant tumor by age fifty than does the average person. People with this mutation are at risk for many types of malignancies, some common and some rare, including ovarian cancer like Veronica's.

Veronica had relatives with multiple cancers at early ages on both sides of her family. If Veronica's mutation was in all of her cells' DNA, that meant it was inheritable and could also affect her two children. I explained to her family that we couldn't tell at this point. Her precise TP53 mutation had not been seen in Li Fraumeni patients in the past. Was that because it was a rare variant? There is much rarer genetic variation that we don't yet know about, the dark matter of the genome. Or was it because relatively few women of her particular Northern European ethnic background have had their genomes deposited in the public databases for us to compare to?

This incidental finding created a great deal of anxiety for Veronica and her family, as if her cancer was now metastasizing to her kindred. And now that this had surfaced, it was important to find out whether she could have passed it on to her kids, since children with Li Fraumeni syndrome need regular screenings because they are in danger of developing cancers in the blood, muscle, brain, and other organs.

For many other cancer genetic risk syndromes that affect adults and not children, we do not test children under the age of eighteen for mutations. Why? Because in these cases, positive results can only cause unnecessary anxiety.

Yet, for Li Fraumeni syndrome, we do test children because there is actionable sequelae: intensive tumor surveillance can reduce misery and save lives. Thus, ironically, when only a tumor carries a TP53 mutation it is not actionable, but when a family does, it is. In fact, it is a disease for which pre-implantation genetic diagnosis is sometimes performed on artificially inseminated embryos so that parents can try to avoid bringing children into the world who are likely to develop cancer.

Veronica's TP53 mutation result came back just as a categorical result: positive and not negative. Yet, one advantage of massively parallel sequencing data is that you can get very quantitative, and this was actionable information. While the quantitative data isn't normally given, we called the testing company and asked what percentage of the DNA sequences carried this mutation. After some delay, they obliged and told us it was precisely 5 percent, suggesting that the mutation was restricted to her tumor and would not be an issue for her family. Li Fraumeni syndrome is a dominant genetic disease, meaning that 50 percent of the DNA in a person with the syndrome carries the mutation, and the other half does not. If Veronica's tumor mutation frequency were right around 50–50, that would be consistent with DNA that had been inherited, and we would have to perform further testing on her normal DNA to evaluate. So in Veronica's case, her family did not have to worry about the Li Fraumeni syndrome. Tragically, though, Veronica did pass away from her cancer and did not benefit from the DNA-based testing.

Another example is Lawrence Andrew Magik, or Larry. Larry was a fifty-three-year-old, well-respected immunologist at a New York–based medical school who was an authority on how antibodies bind to their targets. He was feeling well and in great shape, biking a hundred miles every weekend in New Jersey. Larry never smoked, drank only socially, took only multivitamins, and had no known unusual carcinogenic chemical or radiation exposure during his scientific career. One fine summer weekend day around dinnertime after one of his bike rides, he unexpectedly developed a persistent upset stomach. At first, he attributed it to eating too much sour cream at a sketchy Mexican restaurant. He went in for an examination and found that blood tests of his liver function were abnormal. Then he started to develop pain in the middle of his back. A CT scan showed a mass in the head of his pancreas. Like Tim, he had a Whipple procedure surgery to remove most of his pancreas. Unfortunately, the pathologist could readily see many cancer cells squeezing into the pancreas's veins and lymphatics—not a good sign at all. During the procedure, eleven of twenty-two lymph nodes were surgically excised. The pathologists found that they carried tumor cells consistent with pancreatic cancer, which meant the cancer had already metastasized outside of the pancreas.

Pancreatic cancer is one of the most feared diagnoses in medicine. Five years from diagnosis, only about eight out of every one hundred souls are still alive. If there were ever an urgent need for better therapies for a disease, pancreatic cancer is it.

Larry got the standard initial therapy, a drug called *gemcitabine* that killed all quickly dividing cells, largely indiscriminately. However, despite this treatment he developed three new tumor metastases in the liver. In addition, he was left susceptible to infection because this drug dramatically lowered his immune system's number of white blood cells. To fight the new growths in his liver, he, like Tim, was walloped by the mule of FOLFIRINOX and suffered profound fatigue that made it hard to get to work in the morning. Despite this, Larry became a do-it-yourself oncologist. He read up on the causes of pancreas cancer and discussed *New England Journal of Medicine* articles with his doctors about the pros and cons of different medicines. As a respected colleague, he became an active partner in his own care.

Larry came to me for cancer genetic testing because of his concern about his new diagnosis and family history. His mother had been through a lot. She was diagnosed with colon cancer in her forties, endometrial cancer in her sixties, and then subsequently, fatal breast cancer in her seventies. Larry's grandmother on his mother's side was also diagnosed with breast cancer in her fifties, as was a cousin with breast cancer in her early thirties.

Genetic testing showed that Larry carried a mutation in BRCA2. The breast and ovarian cancer risk genes BRCA1 and BRCA2 are among the most infamous causes of genetic disease. Celebrities such as Angelina Jolie, Sharon Osbourne, and Melissa Etheridge, among others, have done a great public service emphasizing the potential benefits to women at high risk of screening for mutations in these genes before cancer develops. Less well known is that men can also suffer from this mutation, and that the same genes can also increase risk for cancers of the pancreas and other organs.

Larry is Ashkenazi Jewish. Approximately one in forty Ashkenazi

Jewish individuals carries a mutation in BRCA1 or 2. This is compared to a carrier frequency of about one in five hundred in the general American population. Of Ashkenazi Jews who carry these mutations, almost nine out of every ten are caused by only three recurrent mutations. The same three common Ashkenazi Jewish BRCA1 and 2 mutations are also seen in Latinas and Latinos, likely because in the fifteenth century some Jews converted to Catholicism, or secretly practiced Judaism as Crypto-Jews, because of the terror of the *auto da fe* from the Spanish Inquisition (as with Gaucher disease, which I discussed earlier, the mutation likely arises as a result of endogamy among Jews living in Europe at that time).

A few more than one in two women with BRCA1 and 2 mutations will develop breast cancer, and about one in six will similarly develop ovarian cancer over the course of their lives. The risk of developing pancreatic cancer is approximately one in twenty, but this affects both men and women.

Seeking a better therapy, Larry read about a newly developed class of drugs in clinical trials called *PARP-inhibitors*. These drugs have minimal side effects compared to the drugs he was taking.

Cells that have mutations in BRCA1 and BRCA2 have trouble repairing breaks in both strands of the double helix in genomic DNA, also known as *double stranded breaks*. PARP1 is an enzyme important for repairing DNA single-strand breaks, or "nicks." When treated with PARP inhibitors, cancer cells with BRCA1 or BRCA2 mutations die much more frequently than do the normal cells surrounding them. Larry searched on clinicaltrials.gov, a useful National Institutes of Health–sponsored website for finding clinical trials in the United States and found a trial he was eligible for using a PARP inhibitor at a hospital in Philadelphia. While he did have to take five pills twice a day, Larry had minimal side effects from the drug, called Rucaparib. Several months later, Larry was doing well and was able to go back to riding his bike on weekends. We are all hopeful that he will continue to stay well.

There is now hope for those with cancers who have poor prognosis. Both Tim and Larry benefited and were very grateful for the quality time they gained from these treatments.

There are still other arrows in the quiver for Tim. In addition to ERBB2, his cancer also expresses high levels of the epidermal growth factor receptor, or EGFR, a growth-promoting protein on the cell surface membrane that is widely expressed for colorectal, lung, and head and neck cancers. There are several drugs readily available to attack it. Since Tim's original ampullary tumor may have mutated after several rounds of chemotherapy, it is likely that new actionable targets will be sought in the cancer cells that have survived this far. After that, given the ultra-high mutation rates in Lynch syndrome tumors and the high number of tumor infiltrating lymphocytes attacking his tumor, several powerful new immunotherapy drugs that rev up the adaptive immune system to fight tumors are potentially options as well.

While these new approaches are not the Holy Grail of a complete cancer cure, they open up a vision of cancer as a chronic disease that can be managed by understanding the root causes of these awful cells and having a battery of diverse medicines to treat them. Although there is still a tremendous distance for the field of oncology to travel, not only from the scientific point of view but also from patients' financial perspective, the ability to characterize cancers genetically gives us a sense of momentum and optimism.

SIX

Madonna and Putto, Mortar and Plaster

Finding a babysitter on short notice in Manhattan during the middle of a weekday can be a difficult proposition, often more challenging than making a doctor's appointment. So I was flattered when Samantha K. Berkeley, unable to get a sitter and running late, kept her appointment to see me in genetics clinic. And I was both delighted and surprised to see that she brought her beautiful, healthy, newly minted six-month-old son, Giorgio, with her to see me.

I have seen many cute babies in my day, but Giorgio has to be among the most cherubic. Smiling, cooing, and bouncing in his techno-laden Maclaren baby stroller, Giorgio waved his outstretched arms like little Donatello putti wings toward his mother, until she picked him up and placed him comfortably in her lap. Giorgio had soft, wispy blond hair, beautiful round blue eyes, and chubby, pinchable cheeks.

But seeing a baby sitting here on this particular mother's lap, intensely focused on trying to twist apart his brightly colored set of interlocked teething rings, left me very confused. Glancing at my clinic schedule, I saw a notation informing me that Samantha was here because she had been diagnosed with the Loeys-Dietz syndrome. I looked up again at mother and child, and indeed Samantha had *the look* of Loeys-Dietz.

In her early thirties, Samantha was over six feet tall and had a slender build with well-toned musculature. Her face was long, with angled cheekbones, widely spaced eyes (which look even wider when she is smiling), and a petite jaw. The roof of her mouth was likely tall and arched. Samantha's arms were long and graceful, and in fact measured longer than her legs. If she stretched out her arms side to side, like in aerobics class, the fingertip-to-fingertip span was wider than her full height. Samantha's fingers and toes were also elegantly slender, gracefully angled, and could bend in ways mine never could, regardless of how much time I spent doing stretching exercises. For example, she could bend her right and left thumbs so that her knuckles extended past the far side of her palms and index fingers.

Geneticists often call this a *Marfanoid habitus*, named for a related genetic disease whose patients have the same general appearance. The late nineteenth-century French pediatrician Antoine Marfan originally described this syndrome. Dr. Marfan was famous for his meticulous skills of observation and physical examination. While most people with Samantha's features do not have genetic diseases, others do carry mutations in more than a dozen genes that can cause a Marfanoid habitus. As you might expect from looking at passersby walking along, say, the streets of Manhattan, different patients have distinctly shuffled combinations of these different traits. But looks can certainly be deceiving and, beneath the surface of the skin, different mutations in the same genes can still cause *the look* of Marfan syndrome.

The look usually is not caused by genetic disease mutations. Rather, it is "just" the appearance resulting from chance comingling of many different medically unimportant genetic variants and complex gene-environmental interactions, including diet and physical exercise.

Many historical figures had *the look*. These range from leaders such as Mary, Queen of Scots (a lanky five foot eleven inches tall in an age known for short stature), ancient Egyptian pharaohs Akhenaten and Tutankhamen, American president Abraham Lincoln, virtuoso violinist Niccolo Paganini, and, on a dark note, even Osama bin Laden.

The latter was six feet five and weighed only about 160 pounds. His long, narrow visage is well known, but the anti-terror surveillance detection algorithms noted his long and slender arms and fingers as well. His health may also have suffered from Marfanoid-related symptoms. Nearsighted, bin Laden had a limp that some have interpreted as indicating curvature of the spine or other bone abnormalities. He also apparently had digestive issues and a well-publicized kidney insufficiency requiring dialysis.[1]

It has been suggested by those interested in homeland security and medical intelligence that bin Laden may have carried the Marfanoid associated Mendelian disease homocysteinuria. Homocysteinuria is caused by mutations in the gene cystathione beta-synthase (CBS) and, more rarely, sulfite oxidase (SUOX). SUOX mutations have appeared specifically in the bin Laden family's native Saudi Arabia. These genes affect metabolism of sulfur-containing amino acids, including homocysteine. Mutations in CBS or SUOX genes can cause multiple health problems. One of these effects is incorrect decoration of extracellular matrix/connective-tissue proteins with sulfur-containing spongy-like padding molecules that can cause the Marfanoid habitus, brittle bones, and, indirectly, personality disorders and even psychotic episodes. Other effects include blood clots injuring the kidney enough to require dialysis. Ironically, if bin Laden did indeed have homocysteinuria, he may have been alive in the days leading to September 11, 2001, only because he had benefitted from precision medicine treatment with pyridoxine, betaine, or folate. It's conceivable that without the benefits of modern Western medicine, the twin towers four miles south of my clinic might still be standing.

Given that the CIA had possession of bin Laden's body and used his DNA to definitively identify him after the raid into his compound in Pakistan that killed him, it is tempting to speculate that the CIA may have performed genetic testing after his death to gain insight into whether he, and blood relatives who are persons-of-interest to the Department of Homeland Security, carried mutations in CBS, SUOX, or any of the other Marfan-related syndromes.

Some speculate that Abraham Lincoln had a genetic disease called *multiple endocrine neoplasia type 2B*, based on his Marfanoid appearance. This disease also is associated with tumors of the thyroid, in the neck. When Lincoln was young, he was generally clean-shaven or had a small beard. When he was older, he had the familiar longer beard, and his neck was usually obscured by a high collar on his shirt. This has led some medical historians to conclude that he may have been hiding something, such as visible benign tumors on his neck.

Small benign tumors that show up as protuberances on the lips can be a physical manifestation of multiple endocrine neoplasia 2b. Indeed, Lincoln had lumps on his lips (as did three of his sons), shown in photographs taken during his lifetime, which led to speculation he had multiple endocrine neoplasia 2b.

In 2009 genetic tests using archaeological techniques were performed on Lincoln artifacts, in particular a piece of a dress purportedly worn by Laura Keene, an actress in *Our American Cousin*, the play that Lincoln attended at Ford's Theatre the night John Wilkes Booth assassinated him. Historians report that Keene cradled Lincoln's head in her lap after the shooting and that her dress was soaked in the president's blood. However, there was no visible blood on the piece of dress remnant fabric and the DNA testing was inconclusive for the mutation.

In today's culture, Marfanoid traits are a sort of yin-yang of genetic inheritance. Many of its features are the object of great envy and sought by modeling agencies, movie directors, sports recruiters, and high school boyfriends or girlfriends. Our society greatly values and promotes physical features of tallness, thinness, and physical flexibility, even when athletic prowess and height are not in the job description. (Chief executive officers of Fortune 1000 companies are about three inches taller than the rest of us on average, as are American presidents and many models and Hollywood movie stars. Men who are at least six feet tall average more than five thousand dollars in additional salary than their shorter counterparts. Flip through the catalogues of Bloomingdales or Saks Fifth Avenue, the website IMDb, or the sites of presidential and gubernatorial primary contenders and you will see what I mean.)

These traits can lead people to careers in modeling and basketball. In the United States, the National Basketball Association screens for some Marfanoid genetic syndromes as part of its pre-professional draft physical. NBA draftees who test positive are then declared ineligible to join the NBA because of the increased risk of rupture of the aorta.

Marfanoid syndromes can also cause various aches, pains, and other problems, including benign heart murmurs and nearsightedness, which, while bothersome, are not disabling or life threatening. In fact, my medical experience is that most people who have Marfanoid syndromes don't even know it, and their primary care physicians often are unable to figure it out because they only have fifteen minutes to focus on their patients' most urgent medical issues.

For a minority, however, the mutations can lead to potentially catastrophic complications. This was Samantha's situation, as DNA testing revealed. The first genetic disease linked to a Marfanoid habitus was, well, Marfan syndrome itself, as originally identified by Dr. Marfan. It is thought to affect about one in ten thousand people. Ironically, the similar external Marfanoid appearance of many of these genetically distinct diseases with dramatically different internal outcomes may actually have bedeviled Marfan himself. His index patient, a five-year-old girl named Gabrielle, in 1896, likely didn't have Marfan syndrome but another connective-tissue disorder called Beal syndrome, or *congenital contractural arachnodactyly* (CCA).[2]

Marfan disease and CCA are caused by mutations in the Fibrillin-1 and Fibrillin-2 genes, respectively. While this might not sound like much difference, it can make a major impact on a person's life. The most serious complication of Marfan syndrome is sudden death from rupture of the aorta, the main blood-vessel thoroughfare through the body. Marfan syndrome affected Florence Hyman, a tall American women's volleyball player who won an Olympic silver medal in 1984. She died suddenly at a 1986 match in Japan from an aortic dissection and was diagnosed with Marfan syndrome only posthumously. Similarly, Jonathan Larson, writer and composer of the hit Broadway musical *Rent*, who may have been under a great deal of blood-pressure-raising stress as the production approached opening night, died from an aortic dissection the day before the show's New York premiere. He too was diagnosed with Marfan disease—again, only after his death.

CCA, in contrast, has a much more benign course and, more than a century later, history shows that Dr. Marfan's original patient with a Marfanoid habitus, Gabrielle, did not have any known heart issues as she grew into a teenager, which was when the last report on her was written. CCA is also known to occur in other mammals, including cattle, where it is known as *fawn calf syndrome*. Fawn calves also have tall stature from unusually long leg bones, but no known sudden-death problems from blood-vessel rupture or other life-shortening complications of their genetic disease.

Our bodies consist of cells and connective tissue. If cells are thought of as bricks, connective tissue is like the mortar that surrounds the bricks. Different types of connective tissue support, separate, and hold together all the body's organs in their proper places. Connective tissues are sometimes also called the *extracellular matrix*, meaning the space between the cells.

Connective tissues are composed of many types of specialized cells. Some are hard like the plaster that supports the walls of a building (e.g., fingernails and bone), while some are soft and squishable (e.g., the nose and ears). The different connective tissues support cells and make up more than half of the body's volume and weight. Hundreds of genes are involved. These genes make proteins with names like collagens, myosins, keratins, and fibrillins, as well as other –ens and –ins. They form intricately woven structures of multiple self-assembling fibers interlacing like Velcro, and sometimes they can even self-repair.

Many additional genes act to decorate these structures with various ingenious hooks and pulleys, as well insulation and waterproofing materials. These touches can make all the difference, since the same gene-encoded proteins decorated differently can bring about the sponginess of the kidney or the hard consistency of the forehead bone.

In past decades, the connective tissues were often thought to be largely inert slaves to their cellular masters. However, we now recognize that most organs are more inclusive and democratic, and that connective tissues have robust back and forth conversations with their cellular and syncytial brethren to work together to maintain harmony and homeostasis. These signals from the connective tissues surrounding the cells help prevent burst blood vessels, broken bones, acute blindness, and other maladies.

Connective tissues also give critical signals to the blood and immune systems, and slow down cancer cells to keep them from spreading to the wrong side of town. Connective tissue is somewhat like what's now being called a "smart home," a house that monitors and controls itself by, say, automatically adjusting thermostats throughout the day for both comfort and energy efficiency, or by controlling appliances or locking and unlocking doors automatically as occupants come and go, or by warning you while you're on a foreign vacation that your water heater at home has just sprung a leak. These interactions can be enabled via a home wireless network interacting with, for example, applications on that smartphone in your pocket. In the animal body, the connective tissue is the original version of this *intranet of things*, constantly reporting back to the surrounding cells.

In fact, for Marfan syndrome, mutations in Fibrillin-1 have two effects. First, they weaken the structural support of fibers in blood vessels, eyes, spine, and other sites. In addition, the same mutations disrupt interactions holding back cell-signaling activity by another protein called *transforming growth factor beta* (TGF-β). Overactive TGF-β weakens the smooth muscle surrounding blood vessels like the aorta, as well as the extracellular matrix mortar holding it. This insight is likely important, since a particular type of blood pressure medicine, called *angiotensin II receptor antagonists*, can reduce TGF-β as well as lower blood pressure. Angiotensin receptor antagonists such as losartan help prevent aortic tearing for individuals with Marfan syndrome.

As you might expect, there are myriad types of connective tissue and hundreds of genes involved in mixing these mortars. As in any situation with lots of moving parts, so many genes at play can cause problems. Consequently, perhaps as many as one out of every ten people have mutations causing more than seventy known various connective-tissue disorders.

That's a big part of why seeing baby Giorgio happily gurgling away on his mother's lap greatly confused me. For those diagnosed with Loeys-Dietz syndrome (named after the two Belgian and American physicians who mapped the gene mutations in the late 1990s for a disorder once called Marfan type 2), pregnancy is to be avoided at all costs, as it can cause potentially fatal complications. Had I become confused and mixed up my patients? Unfortunately, with clinic overbooking (not to mention network syncing) things can get kind of wild, and this wouldn't have been the first time.

Even although this disorder is rare, affecting only about one in a million people, there are at least four genes whose mutation can cause Loeys-Dietz syndrome. All play important roles in affecting the cellular infrastructure and command and control for properly producing connective plasters collectively given the inelegant rubric of *transforming growth factor beta*[3] and *bone morphogenic protein* pathways, cousins to Fibrilin-1, which is linked to Marfan syndrome.

When one or more of these little genes can't get the plaster to harden to just the right consistency, big problems can occur in the absolutely most inconvenient places. That's precisely why the aorta can be at so much risk. The aorta is the largest circulatory thoroughfare in the human body. It is the main conduit from the heart to every organ of the body—the Broadway of blood, so to speak. The heart pumps a surge of blood every second or so. In patients with Marfan disease, Loeys-Dietz syndrome, and other connective-tissue disorders, the connective-tissue plasters holding together the large pipe that is the aorta can come unglued at the seams with rapid and catastrophic consequences.

Specifically, for those with Loeys-Dietz syndrome, another organ at great risk is the uterus. During pregnancy, the uterus fills up with the baby and fluids and stretches out visibly. If a mother has Loeys-Dietz syndrome, the connective tissues that surround the baby and uterus can rupture, again with potentially horrific complications for both her and her not-yet-born child.

Samantha learned she had Loeys-Dietz syndrome because of her father, who not only had *the look* but also another indication: the progressive enlargement of his aorta could be seen on ultrasound and CT scans, showing that the lining of the aortic pipe was being stressed, a harbinger of bigger problems and the potential need for surgically reinforcing the arch of the aorta with lifesaving plastic mesh. The aortic arch is the place into which the heart pumps directly and so undergoes the most pressure stress. Plastic mesh is like a reinforcing cladding for a leaky pipe that's straining under the pressure of the liquid streaming through it.

The pressure problem in the aorta gets worse because of a vicious circle. Just as blowing too much air in a balloon makes the rubber thin and tense, when a blood-vessel wall bulges out, it becomes weaker at that spot, making it susceptible to further weakening with even small increases in blood flow and pressure. The medical term for this is an *aneurysm*. When an aortic aneurysm reaches a certain size, the dilated portion of the blood-vessel wall is in danger of tearing, or *dissection*, meaning the blood vessel is separating from its covering. This can cause sudden death if the aorta starts to leak. (This malady received national attention when the actor John Ritter, son of cowboy singer Tex Ritter and star of popular television programs such as *Three's Company*, died suddenly, in 2003, of the effects of an aortic aneurysm and dissection at only fifty-four years of age.) Important lifestyle changes to reduce the risk of aortic dissection include keeping cholesterol levels in the low normal range and avoiding smoking.

Unfortunately, aneurysms often don't cause symptoms until they become quite large. When they do become large enough, they can present as severe chest or back pain (depending on which part of the aorta is dissecting away from its connective tissue), dizziness (because not enough blood gets to the head), heart palpitations, or shortness of breath (because not enough blood gets to other organs).

Samantha's father also had experienced three separate episodes of retinal detachment. Detachment of the retina can occur when the connective tissue holding the eyeball in place becomes worn and frayed from just plain living, even if there is no trauma, such as a fall or car accident. It can show up as a sudden flash of light in one eye, as the mechanical stretching of eye photoreceptor cells causing electrical impulses is misinterpreted by the brain as something like a barrage of paparazzi camera flashbulbs.

Today, retinal detachment usually is readily fixable with laser surgery. As with most people with Loeys-Dietz, Samantha's father didn't have any permanent vision loss from the detachment. When this happens once, it typically doesn't attract so much interest from geneticists, because a retinal detachment can happen to anyone. Twice is definitely concerning but could also just be bad luck. However, thrice is almost assuredly anti-charmed and unusual enough that Samantha's father become involved in a research study on Marfan and related genetic disorders. As part of this research study, her father was found to carry a mutation in the least common gene causing Loeys-Dietz syndrome, SMAD3.[4]

Thus, by carrying a mutation not ever seen in any other family in the world in the rarest gene causing a very rare genetic syndrome, lightning had struck Samantha's family two and a half times. Practically, this meant that there just didn't exist much data one way or the other to predict with any precision the likely medical problems Samantha might confront.

We could get at least some clues from her own family history. Samantha told me that her grandmother on her father's side also had the Marfanoid *look*. Additionally, her grandmother had two retinal detachments, so it was likely that she too carried the same SMAD3 mutation. Yet, the grandmother had lived into her sixties, had died of unrelated causes, and never had significant visual loss or aortic or other clearly related problems. More important, the grandmother's connective tissues had apparently not read the textbook on standard problems with Loeys-Dietz: she had had two uncomplicated pregnancies and had borne two healthy sons. However, the grandmother's daughter (Samantha's aunt) did have signs of stress in her aorta, and this was already being monitored. Samantha was tested for the same mutation that her father carried and was found also to have the same SMAD3 mutation.

Samantha is extremely intelligent as well as a highly quantitative thinker, with PhD in the hard sciences from a top-tier university. It is always fun as a doctor when your patients are clearly smarter than you are and can closely partner in their care. Samantha clearly understood her condition and knew a lot about IVF and PGD.

Now, for Loeys-Dietz, there is an additional question of whether or not to use a surrogate mother. This means that if Samantha wished to have another child, her fertilized eggs might be implanted into another woman's uterus to avoid the risk Samantha could otherwise face of a ruptured aorta or uterus.

Despite the fact that these techniques are now well developed, there were also less tangible issues for Samantha to consider. Motherhood was more than a concept to her, something very personal and important. She had chosen to defer her career after all her work in graduate school to focus for the time being on raising her child, or children. Samantha was very concerned about how all this technology would affect the intimacy of her bond with her children during her pregnancy. The genetic technologies would take out "a lot of fun and enjoyment out of the whole experience." Risk calculations could be part of the equation. However, the value of the mother–child bond doesn't distill well into fractional risk computations. Also, since Giorgio had none of the physical signs of Loeys-Dietz syndrome, Samantha had deferred testing him as a child.

When I was in medical school years ago, uterine rupture was thought be an important part of Marfan syndrome. However, now with our progress in human genetics and the ability to split diseases into molecularly defined sub-entities because we know the causative gene mutations underlying them, the data show that this complication is really mostly a feature of Loeys-Dietz (known as Marfan type 2 before we knew the specific genes involved in both diseases that superficially look the same). This distinction allows people with either of these similar-looking disorders to understand more precisely the risks involved.

However, there is still a lot we don't know; in a case like this, we have extremely limited statistical evidence about what will happen to a patient. Because Samantha carried a mutation in the rarest gene causing a rare syndrome, her family's precise mutation had not been documented in the medical literature. The diagnosis had to be extrapolated from the literature about other Loeys-Dietz-causing mutations in nearby amino acids. In her own family history, there were examples of people for whom the gene mutation caused trouble. Yet, there was also the example of her grandmother, who had two healthy boys and a normal lifespan for her era.

To a certain degree, as we try to understand and predict problems, genetic testing competes with family medical history. Additionally, in Samantha's case we were able to do diagnostic imaging studies of Samantha's aorta, which showed no signs of enlargement at all, and therefore had lower risk of becoming problematic during her pregnancy.

Samantha had had a successful and uncomplicated pregnancy with Giorgio and was contemplating having another child. She struggles still with how and whether to use the powerful molecular tools we have at our disposal to reduce her risk. There is no right or wrong answer to this conundrum. The personal preferences in the genome equation overwhelm the hard data inputs. I told Samantha that if it were me or my family, based on what I knew about Loeys-Dietz syndrome, I would likely use IVF and PGD to avoid have a child carrying this mutation. However, my role is not to make Samantha's decision for her. My job is to inform her as fully as I can about what powerful technological tools are available to help her. Her own decisions will help determine her destiny as much as her genome will. Samantha, her husband, and Giorgio are all doing well.

Samantha originally told me she had decided to have another child naturally. However, when I called her later on to see how they were all getting on, Samantha was leaning now more toward ARF and PGD—not for her own sake, but for the sake of Giorgio and his future sibling. She had begun to have issues with her heart because of changes in its mortar and plaster connective tissue from Loeys-Dietz, her first complications from the disease. This experience had made her realize how important it was that she protect herself as well as her next child in order to stay healthy and be able to raise her wonderful children.

SEVEN

The DIY Genome

I had so many e-mail exchanges and phone conversations with Lutz and Genevieve Ringolskij before they finally came to my clinic on a cold winter's day that I felt I had already met them. A successful New York power couple, both had doctoral degrees in their respective specialties. Lutz was a veteran investment manager at an internationally known hedge fund and experienced with quantitative analyses. In particular, he was an expert on risk management. Genevieve was an academic. She taught American and European history to undergraduates at a well-respected private school in New York City. Together, outside of their professional commitments, they spent time together deeply involved in local New York social justice causes and philanthropy, including teaching practical financial literacy to lower-income residents of Spanish Harlem.

Although both of them had some chronic medical issues, in general they were quite healthy, at least by my clinic's standards. Genevieve had ulcerative colitis, a form of inflammatory bowel disease, but no family history of the disease. While this disease can develop in infants, Genevieve did not experience any symptoms until she was an adult. Fortunately for everyone, my notes had a lot more blank space than usual after summarizing patients' family and medical histories. This kind of writer's block is a good thing for everyone involved in genetics clinic. Both Lutz and Genevieve were thankful for their good health and wanted to maintain it.

Their chief clinical complaint to me? Wellness. They had come to see a certified genetic counselor—named Eleanor—and me to find out what they could learn from their genomes to maintain their good health for a few more decades and inform disease prevention for their two thriving tween and teenage children. They did valuable work to help less fortunate New Yorkers and wanted to see this important legacy continue with their continued leadership. This was long-term value investing and a strong hedge against medical crises. They were also interested in self-knowledge—about their ethnic origins and other non-medically useful information to bring their family histories into the twenty-first century.

They understood that they were way ahead of the curve, that medical insurance would not cover wellness genome sequencing, and that they would have to pay cash on the barrelhead for this. In sum, they wanted to have what thousands of healthy (albeit mostly wealthy as well) Americans had already had in their medical files. I like it when my patients are smarter than I am, as was the case here. They came to me after many hours of research and reflection on the topic. Lutz and Genevieve were well informed of the risks, benefits, and limitations on whole genome sequencing.

Being data driven, they also asked for their primary DNA sequence files on a hard drive so that they could personally spend time re-analyzing their sequence data in the future as new disease genes were discovered and old ways of thinking were upended and updated.

The genetic counselor and I met with them for more than an hour. We discussed at length the limitations of genomic knowledge and the existence of genetic variants whose significance we can't readily interpret at present, the potential risks that could arise from incidental findings, and the rare but possible outcome of future genetic discrimination, among other issues. Lutz and Genevieve each separately signed three informed genetic consent forms: one for the hospital, one for the testing laboratory that would do the sequencing, and one for an educational research study to better understand the risk-benefit ratio of genetic testing and whole exome sequencing. There are about 180,000 exons in the 22,000 or so genes, and most of the mutations that we can identify and understand are located in the exons. The exome is the 2 percent of the human genome that makes proteins thought to be the most important products of the genome. Sequencing the exome in general is considered an effective technology to detect the DNA sequence changes that we can clearly interpret as being mutations causing genetic maladies.

As it turned out, their quest for self-knowledge stopped dead in its tracks right there. In New York State, genetic tests can only be ordered for patients if the testing has been pre-approved by the New York State Department of Health. As whole exome and genome sequencing were relatively new genetic tests, becoming available in other states only a couple of years earlier, there were no New York State–approved testing laboratories to order from. Eleanor and I applied for a testing exemption for Lutz and Genevieve given their great intellects and deep understanding of the risks and benefits of wellness exome sequencing. Following instructions, we faxed over the appeal form to allow us to use the molecular laboratory at Baylor College of Medicine in Houston, Texas, which had performed several thousand of these clinical studies with no apparent quality assurance issues raised, and waited. About ten days passed. Then we then received a fax from Albany. The New York State Department of Health had rejected Lutz and Genevieve's requests for whole genome sequencing and for whole exome sequencing. As far as their having a copy on a hard drive of their own genetic code . . . As they say in Brooklyn, it was an emphatic *fuggedaboudit*.

I personally called the state Department of Health to appeal the decision. I explained these were well educated and intelligent patients who had genetic counseling from two board-certified genetic professionals and the intention of using this information for their long-term health, and that we were following guidelines from the American College of Medical Genetics and the American Society for Genetic Counselors. The Department of Health summarily re-rejected us. The reason: Lutz and Genevieve were too healthy for full genome sequencing studies.

In the only example of genetic discrimination that I have personally been involved with as a clinician in my eighteen years of practicing as a geneticist (or really reverse-genetic discrimination in this case), New York State government determined that exome analysis testing at that time should only be performed for persons who have a strong personal or family medical history of a particular genetic disease. The Empire State determined that, absent that sort of medical history, genetic information was too potentially risky and dangerous, and that weighed on Hippocrates's first-do-no-harm scales, the potential benefits of Lutz and Genevieve's genetic information was outweighed decisively by the potential risks to both themselves and the public health of the commonwealth. (I'll discuss some of the public health concerns about widespread sequencing of the genomes of healthy individuals at length in the next chapter, "Generation XX/XY.")

Meanwhile, they could legally have their genomes sequenced and receive all their primary data files in any other state in the country, as well as in Canada. Most states probably allow such testing because the American College of Medical Genetics and Genomics, the leading professional organization for clinical genetics, established guidelines in 2012 that wellness genetic screening was acceptable if healthy persons were appropriately counseled by genetic health professionals (such as was the situation for Lutz and Genevieve) and signed consents acknowledging that they thoroughly understood potential risks.[1]

Should each resident of New York have to carry a card with them warning that knowing their genetic information could be dangerous to their health, I wondered, like the surgeon general's warning label on a pack of cigarettes? Since they can get their DNA sequenced by simply crossing into New Jersey, does their DNA change its properties in the middle of the George Washington Bridge when people drive over the Hudson River?

Or, more plainly stated, whose genome is it anyway? Should healthy people have the right to access their own DNA sequence? Or is this too powerful and potentially toxic knowledge for even well-educated, thoughtful people like Lutz and Genevieve to carry, lest they turn into pillars of salt from exposure?

Genetic exceptionalism is the term describing the opinion that genetic information is distinctively unique and must therefore be treated differently from all other types of medical information. Historically, I believe this echoes the long-discredited arguments for *vitalism*.

Vitalism was the philosophical belief that living organisms are fundamentally different from non-living things because they contain *élan vital*, a special force with no physical basis, sometimes equated with the human soul, sometimes not. Vitalists predicted that organic molecules could not be synthesized from inorganic components, but this was disproven in 1828 when Friedrich Wohler synthesized the main chemical in urine, urea, from inorganic chemicals. Now, two centuries later, must we do the same for DNA and dispel the notion that DNA has magical properties that distinguish genetic tests from all other clinical medicine diagnostics?

Major concerns about the public health implications of DNA testing include how patients will react to *incidental* or *secondary findings*, defined as "the results of a deliberate search for pathogenic or likely pathogenic alterations in genes that are not apparently relevant to a diagnostic indication for which the sequencing test was ordered."[2]

Geneticists approach secondary findings differently from other medical testing. For example, when a patient has a CT of the chest, the radiologist doesn't have to ask before the test "We are doing a CT scan of your lungs. Now, what about the bones? If we find something wrong in the bones, do you want to know about it or for us not to tell you? OK, please sign your consent here. Now, what about the muscles? Alright, now sign again down here. What if we find an aortic aneurysm that could burst? Do you want to know about it?" And so on. Yet, as a New York–based geneticist, I am legally and ethically bound to go this route when my specialty's part of the body, the DNA, is studied.

I argue that DNA does not have magical properties, that it is a powerful source of information, but not one fundamentally different from other types of medical data streams. While perhaps idiosyncratic, it is not categorically unique. My own experiences inform me that people who are properly counseled to understand the risks, benefits, and limitations of genome sequencing should be able to access their own DNA.

There is a long and distinguished history in medicine of self-experimentation. In 1931, Werner Forssmann inserted a catheter into his forearm vein, used X-ray fluoroscopy to guide it up into his neck and down into the right atrium of his heart, went upstairs in the hospital elevator, and took an X-ray picture of it. For this daring feat, Forssmann won the Nobel Prize in Physiology and Medicine. In 1950, William J. Harrington received a blood transfusion from a patient with low platelets and provided evidence of autoimmune antibodies in that patient, whose antibodies then lowered Harrington's own platelets. On the darker side, in 1885 Daniel Carrión infected himself with the bacteria *Bartonella* in order to test whether this specific type of bug was truly making people ill (indeed it was), but consequently died from the disease several weeks later.

Therefore, while I did not (by New York State standards anyway) have a compelling medical reason to sequence my own genome, banking on the examples of Forssmann and Harrington (less so on Carrion), I decided not long after my patients' application was rejected by the state to have my own whole genome sequenced. I did so to help better understand the risks and benefits of this procedure, and also to satisfy my deep personal curiosity about my own instruction model.

First, since I too was in relatively good health (albeit twenty pounds heavier than I would like), I would have to find a genetic professional outside of New York State to have my genome legally sequenced. I decided to ask Henry Lynch, who discovered the Lynch syndrome, to counsel me regarding my personal genome. Henry has been counseling patients for genetic testing for decades. He had also been a close family friend and a professional colleague of my father, who was a gastroenterologist. In the 1970s, Henry and my father had even cruised the American Midwest together in a specially equipped RV van, performing no-cost screening colonoscopies for rural families with strong family histories of gastrointestinal cancer but without good access to medical care. He had known me since I was in elementary school.

Henry is a true geneticist through and through. I remember one story he told me about when he and Ramon Fusaro, a dermatologist, had originally described the FAMMM syndrome in the early 1980s. This is a genetic syndrome of frequent pre-cancerous moles and melanomas. Henry and Ramon were on a plane going to Alabama to study a family with the FAMMM. Henry noticed the flight attendant had atypical moles that were on her neck and on her arms, but not on her face (this is typical for the FAMMM because it spares the face, for reasons still unknown). It was early in the morning, and there were very few passengers on this flight. With true professional sincerity, Henry and Ramon asked the flight attendant if they could do a brief physical exam. In the back of the airplane, the flight attendant took off her blouse. Lo and behold, all over her chest and back were the atypical moles of FAMMM.

While some might have viewed this situation as prurient, their medical and public health intentions were completely professional. After they landed, the flight attendant immediately went to a dermatologist.

"Of course, Steven," Henry said. "I'm happy to help." He paused. "Is there something in particular that you are concerned about? I've known your family for decades, and I didn't think there was any strong genetic disease risk running down your tree."

"No," I readily admitted. "I am an Anxious Alex, one of the worried well. I want to use my genome to help keep me healthy into my old age." I paused. "I also want to just know—the knowledge imperative."

Henry laughed. "Well, it's kind of like that thing with stepping on the glass at Jewish weddings, right? Once you smash it into a thousand pieces, it's done and cannot be undone. Maybe it might end up distracting you from your job and your family." Henry advised me not to rush into having my genome sequence and sleep on it a little more. This take-it-slow approach with multiple discussions is often the norm when counseling for highly penetrant, unactionable diseases like Huntington disease.

Henry's pause-and-think-about-this approach did make me hesitate. Genetic testing is really competing against a more traditional, cost-free way of gaining medical knowledge: one's family and personal history. Would I really learn anything useful I didn't know before? Should I instead just take the money and go to San Diego with my family on vacation? But that would be over in a few days. In the age of medical tourism, where Americans go to other countries for procedures on the cheap, the virtual medical tourism of my genome is something I would always have. This would allow me to bond genetically with my present relatives and all our ancestors for the past two hundred thousand years.

From my personal and family history, I knew that the chances I would find something clearly actionable for me were relatively low. In genetic-risk counseling, this is sometimes referred to the *pre-test probability*. In general, for someone like myself with boring clinical histories, the chances of identifying one or more actionable genetic mutations were about 1 to 3 percent. The personal knowledge data was something that I valued highly, allowing me to investigate my personal and familial genetic origins. Then there was the potential benefit to my children if they could be tested for recessive risk genes that I carry. On the down side, would I become anxious if I carried a variant of uncertain significance in an important disease gene? As Henry Lynch warned, this information might worry me without my being able to take any preventative action. Would this make me anxious or depressed to the point that it interfered with my enjoyment of life?

However, we don't need to have our genome analyzed all at once; we have the option to pick and choose which categories of genes, or even individual genes, we want to hear back about and which we don't. For example, James Watson, co-discoverer of DNA, had his genome posted on the Internet, but he chose not to be informed about potential risk variants for Alzheimer disease and asked to have his information removed from the data reported back to him. That was his personal choice.

For me, perhaps naively, I not only decided to proceed—I chose the whole-genome approach. However, I recognize that many people may choose not to know about gene mutations that are not actionable because knowing could lead to unnecessary anxiety or depression.

Thus, doctor became patient. The next time I visited Henry at his clinic in Nebraska, I had my blood drawn, read my genetic consent form, and sent the blood sample off by Federal Express, at room temperature, since DNA is generally very stable.

While in almost twenty years I had never had any patients who complained to me about genetic discrimination (Lutz, Genevieve, and New York State notwithstanding), and logically I understood this was very rare, now that *I was the patient* I admit I started to wring my hands. I reread the Genetic Information Nondiscrimination Act (GINA) several times, and carefully. If my test did reveal a gene mutation with a poor prognosis (I forgot my house keys the other day. Did this mean I could be developing premature dementia?), GINA protected me from being fired for this reason by my employer. Similarly, I was protected from losing my medical insurance as a result of genetic discrimination. As some patients do, I thought about whether or not I should increase my life insurance coverage and buy additional long-term-care insurance policy. However, considering my pre-test probable risk to be in the low single digits for a mutation of concern (perhaps over optimistically considering myself more like Forssmann and Harrington than Carrion), and given my eagerness for genetic self-knowledge, I went ahead anyway.

Months went by, and sequencers kept on sequencing. Finally, I was united with my doppelgänger of a genome. On a small, silver hard drive almost as thin as my birth certificate, there was my personal instruction manual. It occurred to me that someday this could be used to clone my organs, or even me. If I needed a new 3D printed kidney or heart, this hard drive contained all my tissue-matching genes, the information was all there, prepared in advance. Maybe in the future someone might want to revive me to write a sequel for this book? Or, more darkly, if I died before my house mortgage was paid off, Fannie Mae might want to clone me back to life and make my clone work it off until the balance was zeroed out.

On the interpretative side, my experience as a genetic professional instructed me to focus not on the greys, but on the blacks and whites: the genetic variants that are clearly benign and those that have strong effects on proteins that when mutated have powerful effects on disease predisposition. The blacks and whites are sometimes referred to as the *Interpretome*. These genes are known to greatly increase the risk of disease and the variants that are highly likely to be mutations. This list ended up being fairly short—probably because I was one of the worried well.

On my maternal side, my mother is Ashkenazi Jewish and had breast cancer in her forties. The cancer was detected early by a careful manual breast examination and was resected before the tumor spread to her lymph nodes or anywhere else. She had a full recovery and no new breast cancers or recurrent metastases since that time, now almost four decades ago. Breast cancer occurs at higher rates in Ashkenazi Jews, and my mother met criteria for most insurers to cover breast cancer risk gene testing. Yet, my mother had never had genetic testing.

"What's done is done," my mother said. "I have enough problems as is. With my luck, who knows what I would find." She shrugged her shoulders. "I don't want to think about genetic problems as well." She kept up with her cancer surveillance screening, but didn't want to take up any more of her mental energy on the topic. She just wanted to continue to live her life.

◆ ◆ ◆

Having grown up as Jewish child during the rise of Nazism and the Holocaust, the whole topic of genetic testing and its potential for stigmatization made her nervous. As her son, my risk of carrying a mutation in one of the BRCA genes that increase the risk of breast, ovarian, and pancreatic cancer was estimated to be between 4 and 13 percent. Could I be at increased risk for pancreatic cancer? Or worse, could I have transmitted a BRCA gene mutation to my daughters?

On my father's side, the men consistently have higher than normal cholesterol, although we don't get early onset strokes, heart attacks, or aortic aneurysms. Could I be a carrier for a mutation in one familial hypercholesterolemia genes such as LDLR, APOB, or PCSK9? Because there are now effective drugs available to treat these diseases, this finding could potentially help maintain my health into old age. Yet, the costs of these drugs can be staggering. Given the overall longevity on my paternal side despite elevated cholesterol, would I end up spending my family's retirement savings on drugs that I really didn't need?

My sequence showed that I carried neither a BRCA mutation nor any of the hypercholesterolemia genes. But one bit of information did take me aback. The report indicated that I carried a clear mutation in a well-understood gene that caused disease when only one copy was defective. The mutation cut a critical protein in half so that it clearly could not perform its function.

The gene was ARID1a. This gene causes a rare genetic malady called *Coffin-Siris syndrome* (CFS). ARID1a helps package DNA into neat, well-organized packages. The packages look like beads on a string wound up into spools. This packing helps make it possible for genes to be switched on and off properly when needed. In fact, this gene's function is so important and essential to all the different types of cells in the body that the impact of its mutation is dramatic.

Coffin-Siris syndrome leads to malformations involving the face and the pinkies of the hand. In a developing fetus with the syndrome, cells on the outside of the hand die. Consequently, the children are born with no pinky fingers and, sometimes, without "little piggy" toes. Similarly, during fetal development of the face, certain features become more pronounced, leaving people with the syndrome with a wide mouth; thick lips; a broad nasal bridge and, on occasion, upturned nostrils and nose-tip; long eyelashes; and thick, dark, arched eyebrows that have the appearance of having been drawn by graphite pencil.

First, I looked at my hands. Ten fingers, also ten toes, all present and accounted for. I counted twice just to be sure. Next, I did immediately what any self-respecting clinical geneticist would do. I picked up my iPad, opened my Face2Gene app, and snapped a selfie.

Face2Gene uses advanced facial recognition algorithms originally designed to pick up terrorists in a crowd to help diagnose individuals who have various genetic diseases that cause specific constellations of facial changes, such as the "penciled" eyebrows with thick lips of CFS.

But the Face2Gene computational software apparently wanted to argue with my DNA that I had Coffin-Siris syndrome, confirming my suspicions that something perhaps had gone wrong in the sequencing or analysis process.

ARID1a mutations also cause cognitive impairment—what in the past was commonly referred to as moderate "mental retardation," in just about all people with the syndrome. Expressive language is more severely affected than receptive language is. Other findings include short stature (somewhat true in my case) and hearing loss. (While my hearing isn't that bad, my wife is convinced I have a very strong listening disorder, perhaps a trait that could be mapped to the Y chromosome of all husbands and boyfriends.) Throw in higher risk for cancer of the liver as well.

So what did this mean? Could I have an ARID1a mutation but by chance never have developed any of the signs and symptoms? It certainly was possible that my particular genetic background included variants in other genes that, in a complex way, could suppress symptoms of the Coffin-Siris syndrome. The genetic technical term for this is *reduced penetrance*, meaning a person carries a mutation but doesn't have obvious symptoms. Still, this mutation could be passed down the family line and cause trouble in future generations. Many gene mutations are known to at times "skip" generations because of reduced penetrance.

I wondered if I could write a scientific selfie-case report about this. But I also had two healthy daughters, neither of whom showed any evidence of Coffin-Siris syndrome, nor did my parents or brother. So my skepticism gauge hit the red zone.

I had seen suspicious findings like this before with other patients. When you look at a diagnostic test result and then at the patient, and the two are not copasetic, the doctor needs to consider whether or not there could be quality control problems or a sample mix-up.

Next, I did what any self-respecting research geneticist would do. I had my blood re-drawn and my DNA re-sequenced to confirm the diagnosis of Coffin-Siris syndrome. This revealed that the original ARID1a mutation finding reported from my whole genome sequencing was simply an error, a false positive result that had only served to distract me, distress me, and lead to additional useless testing and wasted money. I now had an increased appreciation for the protective concerns of the New York State Department of Health and realized that maybe I had been too hasty in my criticism of them.

My experience brings up an important point for the worried well who want to have three billion tests done simultaneously when they have their genome sequenced.

When a doctor prescribes a medicine in the United States, it has been tested and vetted meticulously before receiving Food and Drug Administration approval. This imprimatur generally guarantees a high degree of quality control in the entire drug development and delivery process. There are currently a handful of FDA-approved genetic tests, and these have demonstrated two important points. The first is that they measure accurately what they claim to measure with clear standards, and the second is that there is good evidence for what is called *clinical utility*. Clinical utility means that the tests have a substantial amount of evidence behind them that it can provide useful information to direct patient care. For example, testing can provide solid data to support a decision as to whether a woman should have a given type of chemotherapy to treat her breast cancer, or not.

In contrast, for most genetic tests, including whole-genome sequencing, approvals come at the level of individual states, where there is much greater variance in quality standards. In general, the only bar to clear in most states is that a test measures accurately what it claims to measure, with no requirements for clinical utility.

Furthermore, the quality-control requirements are often less stringent. This is likely in part why New York State was early to ban direct-to-consumer genetic tests (those that do not require a genetic professional and genetic counseling but can be ordered over the Internet), such as from the company 23andMe, a direct-to-consumer genetic-testing company. Clients send in a saliva sample, DNA is extracted, and their genetic-testing data, for more than ninety conditions ranging from baldness to breast cancer risk, are posted in a password-protected file on the company's website. In 2008, *Time* magazine named 23andMe the Invention of the Year. In 2013, the FDA sent a cease-and-desist letter to 23andMe and similar companies for direct-to-consumer genetic testing in the United States. However, in 2015, the company did get approved a direct-to-consumer test for a single disorder, a rare recessive cancer risk disease called Bloom syndrome.

There is often not much transparency in providing quality-control statistics to genetic professionals who order these tests and their patients. For example, how deep and complete is the sequencing coverage, as less coverage is more prone to miscalling mutations? What specific algorithm is used to distinguish classification of a genetic variant as benign versus deleterious? In some cases, cross-contamination from other patients' DNA run on the same machines can cause problems as well, and I have seen this arise from at least one genetic-testing company that should have done better.

There are high-quality genomes, and there are discount genomes. I had bought a non-refundable economy F-class genome, and I got what I paid for. When I delved into the actual data and squinted at it, I found that some of my twenty-two-thousand-plus genes didn't really have enough sequencing reads for adequate coverage to detect mutations. When a gene's sequencing is scant, the software sometime calls a mutation when one isn't there. When the quality control isn't so great, this kind of inadequate data can just get pushed through the pipeline.

For me, ARID1a was one of these genes. While the test had covered my whole genome, the sequence coverage was about thirty-fold. The means that for each position in the genome, the average number of sequencing reads is about thirty. However, the range of sequencing-read depth goes from zero in some positions (most of which are not interesting and so are essentially ignored) to more than one hundred because of technical aspects in the preparation of genome sequencing "libraries." As it happened, at the specific ARID1a position, by chance the sequence coverage was only eight. For each sequencing read, there can be errors, mis-incorporation of the wrong base. The computational analysis program uses the average of many sequencing reads to "call" a variant at each position. Unfortunately, the computer analysis program apparently had a hard time with this, and so I got a false positive finding of a mutation that wasn't really there, a computational ghost.

In fact, I found at least a dozen of my genes that similarly had poor sequence coverage over some of the exons and had similar errors, although the specific positions in the genome and predicted amino acid changes weren't important for disease. For example, I had predicted mutations in genes like Actin B that have never been implicated in any human disease, and here were false positives from poor sequence coverage. Still, if these low sequence-read coverage areas were over genes that are clinically important, my genome sequence could easily have missed important findings. Maybe New York State's paternalism wasn't as misguided as I had originally concluded. Garbage in, garbage out, as our computer scientist brethren like to say.

My first genome cost about $2,750. Next, I had my genome sequenced by another laboratory, with higher quality-control standards, for about $5,000. This version came on an iPad with a proprietary genome browser app. I hand checked the results before looking at any of my genome findings. The overall coverage of the genome was much better (more than 40 times better), and I could identify various additional technical indicators that this sequence really was me. It was the genome of a male (check). It also had scores of genetic variants that known to occur at higher rates in Ashkenazi Jewish persons, consistent with my ethnicity.

Overall, I had slightly under four million genetic variants identified. Of this large number, only a handful had been even remotely suggested to relate to clinically important medical conditions. Most of the rest were just a very long list of genetic changes of no known significance, at least at the present time. Scrolling through this long list, I just had to shrug my shoulders and hope that in my lifetime the clinical genetics community would be able to provide more insight into how to interpret which of these, if any, had any meaning other than cocktail party discussion chatter.

About twenty-five thousand of these genetic variants occur in my exome. Consistent with the genetic recipe for *Homo sapiens*, I had about eighty genes where in one of the two copies the genetic changes were predicted to prevent the gene from making any protein at all by introducing what are called *stop codons*. When the messenger RNA made by the gene is translated into protein, stop codons cause only part of the protein to be made. Using the word "genome" to illustrate, it would be like introducing a misspelling that only made the word fragment "geno." Similarly, I had about thirty genes where the stop codons were mutated and the proteins were longer than expected, as is, for example, the word "genome" had become the neologism "genomeuat." However, the important point to me is that none of these longer or shorter gene proteins is known to be medically important at present. Thus, for now they are merely genetic variants of curiosity. For example, intriguingly, I can't smell many sulfurous odors, and I also have several olfactory (smelling) receptor genes that are disrupted. So perhaps I should sign up for research studies on the genetics of smell, or start a company on making new chemicals to block bad odors. But otherwise, this information is not medically relevant, at least for the time being.

Neanderthals are an extinct species of humans whose DNA differs from today's *Homo sapiens* in about one out of every three hundred or so positions in the genome. With regard to personal gnosis, my Neanderthal DNA content was a little on the high side of average, slightly less than 2 percent. This suggested some possible distant Neanderthal relatives in my family about fifty thousand years ago.

The long list of genetic variants included those in skin proteins called keratins that may make skin, hair, and nails more suited for colder environments. Thousands of sequence variants confirmed what my mother had been telling me all along, that I am Ashkenazi Jewish. I am apparently was not one of the surprisingly large number of males descended from Genghis Khan, who lived about thirty generations ago. DNA lineage studies suggest that he and his male heirs impregnated thousands of women as the Mongol horde crossed from Asia to Europe, and that his genetic information is carried by about 0.5 percent of people on the face of the earth today. (However, we don't know for certain that the DNA in question it Khan's, since we have no samples of the original, only that the DNA in question seems to have spread in the same pattern as known movements of Khan and the Mongols across Eurasia.)

I have many known, non-medically relevant genetic variants that have very weak effects. For example, my genome predicts I produce slightly more earwax than average. I also have variants consistent with my brown eyes and brown hair. I will save these bits of personal information for cocktail party talk.

However, when I focused on unambiguous mutations in genes that clearly cause disease, I found that I didn't have any of these. My first response was a modicum of frustration that I had spent all this time, money, and effort yet found nothing explicitly useful to maintain wellness. Of course, I understand that I should be happy about these results. Not having high-risk mutations for cancer or neurological disease is clearly a positive outcome.

Looking to the future, some of the results could come in handy someday if I need certain prescription drugs. If I ever got hepatitis C and needed treatment, I carry a variant in a gene called IL-28B that predicts that if exposed to the hepatitis C virus, I would naturally be more likely to fight off this infection, eliminate the virus, and resist getting hepatitis. This would mean that I would not need treatment. Additionally, this variant predicts that if I did develop a hepatitis C infection after exposure, I would be more likely to be cured if treated with the drugs interferon and ribavirin. This is potentially useful information, as some drugs have troublesome side effects, including aching like you have the worst flu in your life. Knowing that I am more likely to benefit from these drugs might help motivate me to suffer the almost inevitable side effects.

Next, I focused on variants that could affect my family. Since my mother had breast cancer, I was especially relieved for my two daughters that I didn't carry any of the now more than a dozen known breast/ovarian cancer risk genes, including the notorious BRCA1 and BRCA2 genes.

However, I did discover that I am a carrier for four recessive diseases that could be screened for in the next generation. This means I have one normal copy and one mutant copy of these genes. Each of my daughters would have a 50–50 chance of inheriting the mutated gene. That would only be a problem for their offspring if they had children with a mate who also carried a mutation in the same gene. While the risk of this is low, with future genetic testing, my daughters and their potential spouses could reduce to zero the probability of having children (my grandchildren) affected by these recessive diseases.

The first recessive mutation I saw was a mutation in the ASPA gene that has been discussed many times in the scientific literature. In children who have two mutated copies of the enzyme *aspartoacylase*, or ASPA, a chemical called *N-acetylaspartic acid* (NAA) builds up all over the body. This is particularly toxic to cells making a protective substance called *myelin*, which is critical to proper function in the nervous system and functions something like insulation on copper electric wires. Children with this disorder, known as *Canavan disease*, develop larger-than-average heads and experience developmental delay, usually before the age of six months. As children get older, they have trouble sitting, walking, or speaking. Many of these children succumb to Canavan disease as teenagers or sooner. The specific mutation I carry is present in fewer than one in ten thousand Caucasian persons, but has been seen at least a dozen times before in children affected with Canavan disease.

The second mutation I saw was in the gene DHCR7, which affects the body's ability to synthesize cholesterol. When two mutated copies are inherited, it causes Smith-Lemli-Opitz syndrome, named after the American, Belgian, and German researchers who discovered it. Smith-Lemli-Opitz is a disease of cholesterol metabolism, which is important for fetal development, particularly of the brain; disruptions in cholesterol metabolism interfere with production of myelin, an essential fatty substance that acts as insulation and keeps different neuronal wires from getting crossed. Smith-Lemli-Opitz also causes intellectual disability, cleft lip, and gender ambiguous genitalia, but not in all children. Many kids with Smith-Lemli-Opitz are unusually aggressive and perform self-striking and flagellation. There are specific disorders that fall along the autism spectrum, including repetitive upper-body undulations and hand flicking. The specific mutation I carry is present as one mutated copy in about 1 to 2 percent of Caucasians. It likely results in a higher rate of fetal demise so that the number of persons born with two mutated copies is less than one in ten thousand.

I also carry a mutation in the NEU1 gene. NEU1 encodes an enzyme called *neuraminidase* that breaks down yet another type of fatty refuse. When both copies are disrupted, it causes a disease called *mucolipidosis*. In mucolipidosis, muscle and brain degenerate, and the liver and the macrophages of the spleen become engorged with fatty garbage that cannot otherwise be broken down, similar to what happens in Gaucher disease. Laboratory experiments have shown that when the NEU1 gene carries this mutation, the protein that is made doesn't fold correctly into the shape it requires to bind this fat and break it down.

My fourth recessive disease gene is ATP7B. This gene operates like a sort of sump pump that removes copper from cells. We all need a small amount of copper in our diet to survive, but too much is toxic. When this pump is clogged, excessive amounts of copper can accumulate in the liver, the brain, and other organs, a syndrome known as *Wilson disease*. In patients with Wilson disease, a characteristic greenish-brown ring can form around the iris of the eye as copper accumulates there. Symptoms include jaundice (yellowing of the skin), fatigue, bloating, depression, mood swings, clumsiness, and trembling of the hands. My particular mutation is carried by about three in every thousand European Americans.[3]

None of these diseases has impacted my health as far as I can tell because they are all recessive diseases, and I am only a carrier. All four are very rare and generally are not screened for in couples attempting to conceive unless they or family members are affected with the related diseases. Yet now, armed with this information, if they so choose, my daughters, nephews, and cousins could be screened for these gene mutations. Future generations of my family can potentially avoid having children with other carriers of these recessive mutations. And yet, discussing this brave new world with potential spouses could be awkward. Would I be willing to perpetrate genetic discrimination and scare away a potential son-in-law because he has a genetic variant that was difficult to interpret and could possibly be a mutation carrier for ATP7B? I hope not. Nonetheless, I anticipate that as more and more early adopters like myself have their genomes sequenced, there will be pressure on prospective spouses to have genetic testing before having children in order to screen out recessive diseases.

In 2013, the US National Institutes of Health announced a new $25 million pilot program to sequence the genomes of hundreds of babies in the United States, aiming to learn more about how ethically sound and useful this testing could turn out to be. "One can imagine a day when every newborn will have their genome sequenced at birth, and it would become a part of the electronic health record that could be used throughout the rest of the child's life both to think about better prevention but also to be more alert to early clinical manifestations of a disease," said Alan Guttmacher, director of the National Institute of Child Health and Human Development, a program sponsor.[4]

An important rationale for whole-genome screening of every infant at birth is that even though most of the four thousand or so recessive genetic diseases are rare, when you add them all up, they affect approximately 2 percent of the population. Much of this pain and suffering falls disproportionately on our most valuable and vulnerable, our children.

MCADD is one of the diseases for which newborns are screened from biochemical analyses of drops of blood taken from their heels shortly after birth. However, parents are not screened for it genetically because it is relatively rare.

Anne Morris, a colleague in the world of genetics, and my friend, is one mother whose family was struck by genetic lightening and who could have potentially benefitted from more global pre-conception genetic screening. Anne conceived a son with a sperm donor. Sperm banks typically screen donors for more common genetic diseases, such as cystic fibrosis, but not for rare recessive diseases.

Both Anne and the donor carried a mutation in the gene ACADM, which causes the recessive disease *medium-chain acyl-CoA dehydrogenase deficiency*, or MCADD. This disorder affects about one in ten thousand births. ACADM is a metabolic gene that helps convert fats into sugar to provide energy for the body. Newborn babies develop low blood sugar, often after periods of being sick with infections and not drinking enough formula or breast milk. When undiagnosed, MCADD is a significant cause of infant sudden death, or seizures after a mild illness. At the same time, other children with MCADD may remain symptomless, provided they don't experience a long period of fasting that otherwise would lower their blood sugar too much. When properly followed by diligent family members and medical professionals, children with MCADD can live long and full lives. In fact, with knowledge of his MCADD diagnosis, Anne and her family were able to keep her son (who is now in his teens) healthy and thriving by meticulously watching his diet: genetic preventative medicine at its best.

Anne started a genetic-testing company for prospective parents and now works diligently as its CEO to help reduce the risk of this happening to other parents and their children because genetic lightning clearly strikes more often than it has to. As with vaccination programs, if everyone were screened for these six hundred or so genes, much pain, anguish, and even expense could be avoided.

I found that I had millions of genetic variants of uncertain significance that aren't clearly understandable as disease-related or not. While they may not help my family or me right now, having them sequenced could turn out to be an investment in the future. This highlights an important point: If you have your genome sequenced, you want to choose a testing laboratory that will continue to update the interpretations of variants as new medical genomic knowledge accumulates over time and then alert you with the information. *Caveat emptor* for those who choose bargain-basement genome sequences.

Getting my genome sequenced and finding out that the probability for a long, healthy life ahead was high left me feeling invigorated and motivated to take better care of myself for the long run. I vowed to exercise more regularly, eat better, lose weight, and save more for a retirement that could be a long one. I also learned a bit more about my heritage and my relationship to humankind. And I'm gratified that my findings may benefit and inform future generations of my family.

I also recognized that my genome might become useful to help other people not related to me. There are so many known unknowns now, the variants of uncertain significance, that the only way to increase understanding of them is if we all to work together and examine genome-wide data in as many people as possible in order to understand which variants cause of disease. I am hopeful that my medical history and genome can be deposited into a databank so that it can help inform other people. I am eager to participate in a large international effort that will, I believe, support medical research and make the world a better place.

Personally, I also gained a better appreciation of the need for genetic privacy. They say you don't have to have had hepatitis to treat a patient with hepatitis, but in the same way that becoming a parent made me a more empathic doctor for pediatric patients and their parents, I believe having my genome sequenced has made me more understanding about concerns about genetic privacy and discrimination. At the Mount Sinai School of Medicine in New York, there is a course for graduate students who study their own genomes, and all board-certified psychiatrists are required to have been a patient in psychodynamic therapy for a least a year. Perhaps medical students and genetic professionals should consider having their genomes sequenced to be more empathic for their patients about genetic medicine.

On a final note, it was early in 2014 that New York's state health department turned down what I thought was a reasonable request by my healthy patients to have their genomes sequenced. In October of that year, the New York Department of Health did change its policy: it now allows whole exome and genome sequencing without state pre-authorization. New Yorkers similarly interested in wellness can now have their exomes or genomes sequenced here without having to pay the extra toll and sit in traffic on the George Washington Bridge.

In any case, in 2013 Lutz and Genevieve ended up going to New Jersey to have their exomes sequenced after being denied by the Empire State. Fortunately, they turned out to have medically boring exomes. In my opinion, genetics tests and laboratories should have a ratings system for consumers, similar to restaurants in the city. An A-grade would indicate FDA-level quality. A B-level test would have good technical quality but not necessarily clinical utility. A C-grade test would indicate lower quality in both spheres, and so on. There is currently an ongoing debate among the FDA, Congress, and the genetic-testing industry regarding how strict regulation of genetic test quality should be in the United States, and where to draw the line between quality control and slowing down the pace of bringing new genetic tests online.

EIGHT

Generation XX/XY

Whole genome sequencing has now become a routine technology in the medical care of citizens in the developed world with a wide range of illnesses, and hundreds of thousands of people have already had their genomes or exomes fully sequenced.

These intellectually intoxicating advances have created great enthusiasm not only among scientists, physicians, and patients but also among policymakers and public health experts. This is the basis for a new paradigm in precision health care (also called *personalized* or *individualized medicine*) that promises to decrease the number of Americans with diabetes, cancer, and other illnesses.

Because we *can* do it, there are now serious policy discussions both in the United States and abroad among academicians, policymakers, and public health experts about whether, in the coming decade or so, we should begin sequencing the genome of every American born baby: Generation XX/XY. Whole genome sequencing at birth would augment or replace current newborn screening that uses a heel prick for a spot of blood followed by tests that focus on a couple dozen diseases that are highly actionable (mostly genetic but also typically including HIV and some other nongenetic diseases). The hope is that whole genome sequencing at birth will give us the ability to better individualize medical care directly from the beginning of life. That could mean having our cake and eating it too: better health from precision medicine from birth, even while improved prevention helps us restrain the raging monster of upward-spiraling health-care costs.

If we were to sequence all those baby genomes, one likely goal beyond precision medicine for those children and the adults they will become would be to reduce the prevalence of at least the worst of the approximately six thousand to eight thousand Mendelian diseases in the population as a whole over time. These maladies collectively cause enormous pain, suffering, and death, as well as consume a disproportionate amount of health-care resources for both public and private sectors alike. For example, it may be possible to detect several rare diseases that disable the immune system and can cause life-threatening infections in children and then treat them with bone marrow transplants. Like fighting fire with fire, you fight genetic disease with genomics, right?

A potential pilot model for trying to reduce suffering and from using whole genome sequencing to detect in children Mendelian genetic diseases is the crusade of a Brooklyn, New York–based international organization called Dor Yeshorim (דור ישרים; Psalms 112:2), which translates as "The Upright Generation." As a case study, the Dor Yeshorim story is tells of great success and possible pitfalls.

In the Book of Job, we are told that Job was a successful patrician living in northeastern Israel with his loving family, including seven sons and three daughters. Then, God tested his pious servant's faith. All of Job's children perished at Satan's hand, even as Job suffered the loss of his comfortable life and property. If this canon of the Old Testament were written in contemporary times, Satan's curses might have included a mention of Tay-Sachs disease, a genetic disease from the deepest pit of hell.

In Tay-Sachs disease, at around six months of age, a thriving beautiful baby boy or girl suddenly can't hold its head upright anymore. Then, the child has trouble swallowing, thus losing nourishment. Next, the child's cognitive and motor skills regress, followed by seizures, paralysis, and blindness. As the disease progresses, a vegetative state ensues, usually followed by death at around age three to four from failure of the nervous system. Milder forms of Tay-Sachs disease can also present in teens with progressively worsening muscle weakness followed by problems with walking, slurred speech, and psychiatric illnesses.

Tay-Sachs was first identified more than a century ago by New York–based neurologist Bernard Sachs, who noticed that the very young children in his practice who were affected by the disease were very often the offspring of Ashkenazi Jews (Sachs was one himself). Tay-Sachs is a recessive Mendelian disorder. This means that for a child to be affected, he or she must inherit two mutated copy of the gene HEXA, one from each parent. Meanwhile, with their dominant HEXA genes intact, the parents have no obvious warning symptoms. The odds of any one of their children inheriting one copy of the recessive gene from each parent, and thus winding up with the disease, is one in four. Thus, for many of these families, particularly those who already had their first children born unaffected by the disease, these events came as a complete and terrible shock. In 1969, Tay-Sachs was first shown to be an inherited genetic metabolic disease, and a screening blood test was developed to detect parents who were carriers. The first pilot screening program was initiated in 1971 in Ashkenazi Jewish residents of greater Baltimore, Maryland. Since then, screening has reduced the number of children afflicted by Tay-Sachs disease by almost 90 percent.

Each ethnic group has its own diseases that are more or less common because of their genetic history. For example, the genetic blood diseases called the *thalassemias* (discussed in the chapter "Altitude Sickness") are more common in persons whose heritage is from Africa or the countries of Southeastern Asia. Overall, different ethnic groups are surprisingly similar genetically across the planet. When averaged over the entire human genome, about nine out of every ten genetic variants found are seen pretty much in any human ethnic group. So, we would expect two people randomly chosen from Europe or Asia to differ genetically by just about one-tenth more than another two individuals chosen at random from Europe or Asia. HEXA gene mutations that cause Tay-Sachs are not specific to the Ashkenazi Jewish ethnic group, but they are more common among this particular population. The precise reasons why this is so are historically not clear.

Rabbi Joseph Ekstein, whose own Job-like trial was watching four of his children die from this disease, founded Dor Yeshorim in 1983. Dor Yeshorim and similar programs are credited with the dramatic reduction in this disease among Orthodox Jews. With more than 150,000 individuals screened, incredibly no children with Tay-Sachs disease have been born to participants in Dor Yeshorim's screening program to date.

Dor Yeshorim started out screening for Tay-Sachs disease alone.

With the identification of genes causing nine additional devastating recessive genetic diseases more common among Ashkenazi Jews, the panel has grown to ten disorders. Orthodox Jewish high school students are tested several years before they would contemplate marriage. They undergo testing using a strategy called participant de-identification. There is a unique alphanumeric code (like a PIN code) linking the blood or saliva donor and his or her DNA. However, a person's name is never directly listed next to genetic results, and neither the teens nor their parents are given the results.

Potential marriages to this day among Orthodox Jewish young men and women are still often still are set up by matchmakers. The matchmakers, who are mostly women called *shadchonim* (although kids often call them busybodies), administer a formal matchmaking system called *shidduch*, a tradition still similar in many ways to matchmaking for Tevye's daughters and their future husbands in the popular musical *Fiddler on the Roof.*

A call to the Dor Yeshorim hotline with the PIN-like codes for both parties will give the potential couple the genetic information they need. Precisely when Orthodox Jewish young men and women call Dor Yeshorim for a "carrier check" is a topic discussed with some delicacy over tea, kosher pastries, and doilies. In principle, the topic of carrier checks can be addressed before the first date. In practice, the matter can be awkward. Some parents might feel that the kids go on too many first dates, and thus end up calling hotline too many times. Others feel that asking for a carrier status check after a first date puts too much stressful pressure on the kids to commit early to a marriage. Worse, requesting a carrier check can be perceived as one young person appearing pushy or too eager to rush into a relationship before everyone feels comfortable. At the very least, even as a non-Orthodox Jewish clinical geneticist (and parent), I can appreciate that discussing genetic diseases before a first date is a real romantic mood killer for the embarrassed kids, who certainly have many other more important topics on their minds. In fact, some parents have to be aware that the need for Dor Yeshorim carrier checking could also serve as an excuse for shy young men and women not to date at all.

As it turns out, in practice, the carrier check is often done around the third date, before things get too serious. When the kids or their parents telephone the Dor Yeshorim hotline, they are told if there's a risk of having an affected child with that particular potential mate.

The Dor Yeshorim organization does not disclose to them which disease genes they carry. Abortion is controversial in Orthodox Jewish law and only permitted in very limited circumstances. However, if the relationship continues, the future newlyweds can use pre-implantation genetic diagnosis to lower their risk of having an affected child. They might also choose to adopt or just take their chances and risk having an affected child.

When a potential genetic carrier problem is detected for a couple, and therefore a relationship is abruptly discontinued, there are fears that the individuals or families involved can be stigmatized as others in the community pick up the signal that a genetic problem might be the reason. The formerly courting kindreds can be stigmatized as "genetic risk carriers" when they subsequently pursue other matches.

Not surprisingly, on the other end of the spectrum, there is also much kvetching and significant ethical concerns raised about the approach used by Dor Yeshorim to screen for genetic diseases. Rabbi Moshe Dovid Tendler, a professor of medical ethics at Yeshiva University in New York City, calls what Dor Yeshorim is doing as "affirming eugenics, the idea that Jews are the repository of bad genes."[1] The concern that Dor Yeshorim is paternalistic has also been raised by others who interpret the organization's intentions as having an Orwellian Big Brother approach to the problem of trying to reduce the number of children born with genetic diseases. Additionally, as described in a letter in the *Jewish Chronicle* regarding Dor Yeshorim's Tay-Sachs screening program, people who are tested and have been shown to be recessive mutation carriers for Tay-Sachs and other diseases have expressed significant feelings of stigmatization, personal devaluation, and psychological distress after testing positive for one of these disorders.[2]

While I share some of the concerns raised about Dor Yeshorim, my own private opinion as a physician and genetic health professional who has cared for many individuals with genetic diseases is that it is a great *mitzvah* (good deed). When I was born in the 1960s, Tay-Sachs disease was more than ten times as common as it is now. I have only seen one child with Tay-Sachs in my entire medical career (the child was of Cajun ancestry, not Jewish, and so not screened by Dor Yeshorim). Much of the gratitude I have for not having seen more of these patients is owed to Dor Yeshorim and similar organizations that have helped increase awareness of the need for screening among non-Orthodox Jews and other ethnic groups.

Now fast-forward to the impending newborn Generation XX/XY. If whole genome screening of all American babies comes to pass, it is highly likely that the incidence of many Mendelian diseases, particularly recessive diseases, could be dramatically reduced, since many carriers would be have information that would lead them to consider PGD, surrogate parenthood, adoption, and other options. Politically, there is perhaps an interesting opportunity for bipartisan consensus here, because screening would likely reduce significantly the number of abortions in the United States each year, as problems that are now detected only after a pregnancy occurs would be reduced.

Yet could widespread genetic testing also carry serious risks along with any benefits? "Infant DNA Tests Speed Diagnosis of Rare Diseases," read a recent *New York Times* headline.[3] How can this be bad in any way, and who could possibly be against genome sequencing newborns as soon as possible? Fear of genetic discrimination and affirmation of eugenics are perhaps the two most important concerns.

The evolution of genetic discrimination, and perhaps the overenthusiasm about the benefits of genetic information, predate the discovery of DNA and can be traced back to the late nineteenth century and the birth of eugenics.

The term *eugenics* was first used by the British scientist Sir Francis Galton in the 1880s. Galton was a brilliant psychologist, sociologist, anthropologist, and overall polymath (he also is credited with making the first weather map) who made seminal discoveries in the field of statistics. His discoveries include key concepts still used today in genetics and many other areas, such as correlation (e.g., children with freckles and red hair correlate with parents who have freckles and red hair).

As a pioneer in the social sciences, Galton was the first to create surveys and questionnaires to collect data in order to analyze various sociological, psychological, and biological characteristics. He is credited with coining with the widely used expression "nature versus nurture" to describe the interaction between genetics and environmental influences. It is therefore not surprising that Galton wanted to apply social scientific techniques to improve social conditions in Great Britain. His book *Hereditary Genius* was among the first to correlate achievement and wealth with family traits. Using sources such as *The Dictionary of Men of the Time*, which were similar to today's "Who's Who" books, Galton noted that a statistically larger than expected fraction of British men distinguished in the arts, sciences, and politics were related by blood, compared to the overall British population. Galton concluded that inherited factors dictated not only height, skin color, and other physical characteristics but also abilities such as intelligence, wealth, social rank, and creativity.

Perhaps not coincidentally, the late nineteenth and early twentieth centuries were a time of breathtaking scientific progress in both the life and physical sciences. This period saw the discovery of the Theory of Relativity in physics and the widespread adoption of revolutionary innovative technologies, including the light bulb, the telegraph, and the automobile. Charles Darwin's theories of evolution (Galton was a cousin of Charles Darwin) were becoming accepted more and more by the mainstream public, and the classic genetic experiments of Gregor Mendel resurfaced after several decades in obscurity. Thomas Morgan and colleagues were establishing Nobel Prize–winning fundamental principles of genetic inheritance in fruit flies. It was also a time of rapidly growing population, intense competition for limited resources, increasing ethnic diversity, and, during what was also the (first) Gilded Age, increasing social inequality.

Galton defined eugenics as the "self-direction of human evolution."[4] Just as engineering is an applied science of physics, eugenics was an applied social science. Its overall goal was translating the dramatic basic scientific advances of that period into a greater good by promoting a higher rate of reproduction of certain people with traits seen as desirable, and, ominously, by reducing the reproduction of people with traits judged as less desirable. Eugenics was also a scientifically mechanistic rationalization of racial and class prejudice.

Eugenics achieved mainstream acceptance in both North America and Europe in the early twentieth century. Its main proposed approach was preventing "undesirables" from having children through sterilization techniques such as hysterectomy and vasectomy. There was even discussion about sterilizing individuals who suffered from complete color blindness. The eugenics movement was not a lunatic fringe but a mainstream social philosophy, often even embraced as "progressive" science-based idea. In 1913, President Theodore Roosevelt wrote, in a letter to the well-known eugenicist Charles Davenport, "Someday we will realize that the prime duty, the inescapable duty, of the good citizen of the right type is to leave his or her blood behind him in the world. Wrong types need not apply."[5]

Long Island, New York's Cold Spring Harbor Laboratory, today a storied genetics mecca that continues to make state-of-the-art advances in our understanding of the field, was actually once an American epicenter of eugenic "research." Its Eugenics Record Office was founded in 1910 with funding from the Carnegie Institute and the Rockefeller family.[6]

Winston Churchill, concerned about how England would be able to compete with the rising power of a larger Germany, championed British eugenics. He proposed using a state-of-the-art technology of that era, X-rays, for efficient sterilization of undesirables, such as those with developmental delay and (often presumed but not necessarily genetic) physical infirmities.[7] Eugenics became an academic discipline at many colleges and universities, which also helped increase its respectability. Notable and respected public figures who supported eugenics included the economist John Maynard Keynes (director of the British Eugenics Society), Linus Pauling, George Bernard Shaw (who discusses eugenic philosophy in his classic play *Man and Superman*), H. G. Wells, Alexander Graham Bell, and Oliver Wendell Holmes.

In the United States, perhaps the most infamous case of eugenics in practice was the matter of legally enforced sterilization of individuals without their consent, which was addressed in the Supreme Court case *Buck v. Bell* in 1927. Carrie Buck's mother, who was African American, lacked an education and was medically diagnosed as "defective" and institutionalized in the Virginia Colony for Epileptics and the Feebleminded. As a teenager, Carrie was raised by a foster family because her mother was institutionalized. A distant relative of her foster family raped her, and she gave birth to a daughter. Whether due to genetic or nongenetic causes, Carrie's daughter did not achieve standard developmental milestones and was diagnosed as "feebleminded" at age six months.[8] Subsequently, Carrie's mother was evaluated by local physicians and diagnosed as "hereditarily defective." Although she also lacked an education, she wanted to eventually marry and have a family, and so resisted her involuntary sterilization in a series of trials and appeals. One of her court-appointed defense attorneys was Aubry Strode, a lawyer who was a known supporter of eugenics, and perhaps not completely unbiased.

In 1927, the case went to the United States Supreme Court. The court ruled against Buck, mandating her sterilization and preventing her from having more children. In the historic *Buck v. Bell* decision, eugenicist and chief justice Oliver Wendell Holmes upheld forced sterilization of family members of kindreds with inherited developmental delay with the infamous quote, "Three generations of imbeciles are enough."[9] In his majority decision, Holmes wrote, "It is better for all the world, if instead of waiting to execute degenerate offspring for crime, or to let them starve for their imbecility, society can prevent those who are manifestly unfit from continuing their kind."

The United States was the international leader at the forefront of the eugenics movement. From the 1900s all the way to 1981, approximately sixty-five thousand American citizens were forcibly sterilized without their consent in the more than thirty states under the eugenics laws. California was in the vanguard of the American eugenics movement, sterilizing an estimated twenty thousand Golden State residents during these decades. In the 1930s, the Adolf Hitler cited the American statewide sterilization program in California as a critical demonstration that large-scale eugenic solutions, including compulsory sterilization of undesirables, could be implemented effectively.

Tragically, Hitler and the Nazis interest in eugenics soon took matters far, far beyond forced sterilization. Citing some of the same reasoning as the eugenicists, by 1939 the Nazi government had put into place a program of what they characterized as "euthanasia" of mentally and physically disabled infants and children. Medical professionals were ordered to register names of children born with severe disabilities, with small teams of physicians deciding which children would be subjected to this so-called mercy death in special "clinics"—more than 5,000 by the end of World War II, in 1945. The killings were soon expanded to disabled and mentally ill adults, with more than 70,000 gassed in carbon monoxide chambers disguised as showers, and another 130,000 killed by techniques ranging from overmedication to starvation.[10] The culmination of what they saw as racial—that is, genetic—cleansing was the Nazi's "Final Solution," the plan to mass-murder virtually all European Jews, homosexuals, and other "undesirables." At the Nuremburg trials, Nazis who committed genocide defended themselves by arguing that the US eugenics movement and Oliver Wendell Holmes's ruling in *Buck v. Bell* was the inspiration for the construction and implementation of the death camps.[11]

The more than 65,000 unfortunate victims of legally enforced sterilization programs in California and other American states included many of the most vulnerable US citizens. The victims were defined as "socially inadequate persons, which included individuals who were feeble-minded, insane, criminalistic, epileptic, inebriate, diseased, blind, deaf, deformed, and dependents such as orphans and the homeless."[12] In 2003, California governor Gray Davis officially apologized for the state's eugenics program and forced sterilizations.[13]

While morally repugnant in humans, selective breeding of animals can direct their evolution and reduce the incidence of undesired or undesirable inherited traits, and such practices have been well appreciated for thousands of years. Farmers and ranchers had long known they could enrich for specific desirable traits in their flocks by controlling breeding. Many considered this inherently obvious. At a time when people and animals lived much closer together than they do today, perhaps this is partly why eugenics achieved such mainstream acceptance. A brew of sound science and immoral state-enforced policies that emphasized the greater benefit to society over the rights of citizens made eugenics particularly chilling.

Thirty-three US states passed eugenics laws between 1900 and 1925, with the last revoked in Oregon only in 1983. In Oregon, eugenics laws were initially passed in 1913, and a state Board of Eugenics was created ten years later. Later renamed the Oregon Board of Social Protection, this agency existed into the presidency of Ronald Reagan, with the last court-mandated forcible sterilization occurring in 1981. From the 1920s until the 1980s, the state of Oregon alone legally enforced the sterilization of 2,648 men and women by castration, vasectomy, tubal ligation, and hysterectomy. These included pediatric residents of reform schools and even teenage girls who were considered nondisabled but morally inferior and labeled "promiscuous." In 2002, Oregon governor John Kitzhaber officially apologized in public for these state-enforced mandatory sterilization procedures, although there has yet to be reparations to the victims of eugenics in that state.[14]

Of the thirty-three US states with eugenics laws, Virginia and North Carolina recently became the first two states to compensate victims of their eugenics programs, paying $25,000 and $50,000, respectively, per individual. Virginia's last legally mandated forcible sterilization was performed, and the program officially ended, in 1979.[15]

One victim who eventually received compensation from the state of Virginia was Louis Reynolds, an African American. In an accident at play when he was thirteen, Reynolds was hit in the head with a rock. A doctor subsequently misdiagnosed Reynolds as a congenital epileptic, genetic and heritable, when in fact his symptoms were a temporary consequence of having been hit. During his hospitalization, the young teen was sterilized by vasectomy without his consent or even his knowledge of the medical procedure while he was under anesthesia. He only discovered it years later after he married and wanted to have children. He would go on to serve in the Marine Corps for three decades, and then become an electrician after retiring from the service. "I think they done me wrong. I couldn't have a family like everybody else does. They took my rights away," Reynolds told the Associated Press in 2015, when he was one of only eleven surviving victims of an estimated seven thousand eugenic sterilizations in Virginia to receive the allocated $25,000 in compensation from the state.[16]

Not much later, when the Human Genome Project was launched at the National Institute of Health's National Human Genome Research Institute in the 1990s, James Watson, its first director, who had been director of the Cold Spring Harbor Laboratory, instituted an ethics program. "In putting ethics so soon into the genome agenda, I was responding to my own personal fear that all too soon critics of the Genome Project would point out that I was a representative of the Cold Spring Harbor Laboratory that once housed the controversial Eugenics Record Office," Watson said. (Later, in 2007, Watson drew plenty of controversy over comments he made to a British newspaper suggesting that black people were less intelligent than white people.)[17]

Today, state-mandated involuntary sterilizations targeting individuals with genetic mutations seem far-fetched and unlikely, but the fear of more subtle types of genetic discrimination has become more common. In 1997 the United Nations Educational, Scientific and Cultural Organization (UNESCO) formulated a policy statement against "discrimination based on genetic information."[18] Still, few clear-cut contemporary examples of documented genetic discrimination have actually emerged. In the journal *Genetics in Medicine*, a 2000 article about patients' perceptions of genetic discrimination stated that "there may be so little actual discrimination that it may not be possible to initiate good test cases," and, "Patients' and clinicians' fear of genetic discrimination greatly exceeds reality."[19]

In 2001, the first case of alleged discrimination was filed by the US Equal Employment Opportunity Commission against the Burlington Northern Santa Fe Railroad. Highly publicized, this incident is both stranger than fiction and well documented in the Federal Register. The Burlington Northern Santa Fe Railroad was formed in 1995. Formed during a corporate rollup of the Atchison, Topeka, Santa Fe, and Burlington Northern Railroads, it currently hauls enough coal to generate roughly 10 percent of the continental United States' electricity. Warren Buffet's Berkshire Hathaway bought the company for $44 billion in 2009. This railroad company has no direct connection to genetic tools, testing, or even health care, and certainly has never had a reputation of being particularly medically, or genomically, sophisticated.

Carpal tunnel syndrome is a common disorder that can cause weakness, pain, and numbness in the vicinity of the thumb. It is most often caused by repetitive motion (like the typing and clicking I am doing right now) or pressure on the wrists. Other common causes include wrist fractures from falling and using an arm to break a fall, playing musical instruments, obesity, and, if you have two X chromosomes, getting pregnant. If a physician takes the time to perform a detailed physical exam, tapping with her finger over the belly of the wrist may cause a shooting pain extending from the wrist to the hand (Tinel's sign). Or, flexing the wrist for about a minute may cause some tingling, discomfort, or weakness (Phalen's test). The most common treatments are splints, ergonomic adjustments, and avoiding repetitive wrist movements. It is a very common cause of work-related injuries involving iterative activities.

I occasionally experience CTS (its medical acronym) when I am on my road bike for a long time and the handlebars are raised so high that my wrists extend back too far, which puts pressure the median nerves. It is unlikely linked to any genetic disease, particularly now that I have had my own genome sequence in hand, and I have discovered a low-tech cure: lowering my handlebars. Regardless, despite the fact that CTS is commonly related to various kinds of repetitive activity, not DNA, it was at the center of the most high-profile genetic discrimination lawsuit in US history.

In 2000, Burlington Northern Santa Fe commenced sequencing the DNA of some twenty of its employees who complained of CTS—*without telling them*—for an orphan genetic disease affecting less than 0.01 percent of the population: *hereditary neuropathy with liability to pressure palsies* (HNPP).[20]

HNPP was described first in 1947 by a Dutch neuropsychiatrist, J. G. Y. De Jong. He reported that, across three generations, a family had "bulb diggers" palsy from planting tulips. Family members also reported weakness of the lower legs after digging potatoes in a kneeling position, usually with symptoms first appearing during the teenage years. This was episodic, coming and going, with transient weakness. When examined by a doctor, these patients' ankle reflexes were often weak or absent completely, but they would later recover. This same disease was also called "slimmer's paralysis," after some younger dieting women started having symptoms after waking up in the morning, presumably from having less anabolized chocolate ice cream and fewer madeleine cookies to cushion their nerves. In another kindred, an eighteen-year-old whose hobby was jumping out of planes reported acute onset of weakness in his left shoulder after parachuting. In yet another, a girl developed a droopy left shoulder after carrying a heavy backpack to school. Oh yes, all of these symptoms could be associated with CTS. In 1999, a family was reported with multiple generations with CTS from HNPP.

HNPP is caused by deletions in a gene encoding a protein important for myelin, the substance that forms a protective coating around nerves. There is no particularly strong association of CTS with the railroad industry. There are no known HNPP cluster hotspots with a high incidence of cases in the western United States, where Burlington Northern's employees work and live. Ordering HNPP testing in workers complaining of CTS made about as much sense as AIDS testing for everyone who caught a cold last winter.[21]

In 2001, the US Equal Employment Opportunity Commission (EEOC) sought a temporary injunction against the railroad from conducting additional tests while it proceeded with an investigation. When it later charged the railroad with violating the Americans with Disabilities Act by performing genetic testing without patient consent, Burlington Northern didn't put up much of a fight. In May 2002, the company signed a mediated settlement with the EEOC, agreeing to essentially all stipulated demands and a $2.2 million penalty. Precisely which individuals in the company were aware of, and authorized, this policy was never clearly articulated.[22] Burlington Northern claimed it performed HNPP genetic testing for this rare disorder to determine why so many of its employees had repetitive motion injuries like CTS. Ironically, the cost of genetic testing to identify an individual with HNPP likely far exceeded the potential cost savings from this search.

Was this just bad medicine by confused, occupational health physicians who never had training in human genetics and were just trying to do their jobs? Was there a mandate from the corner office to lower expensive medical and workers' comp claims at all costs, by whatever means necessary? We may well never know the true answer. Since, apparently, not a single Burlington Northern worker tested positive in this covert DNA testing, one can argue that this wasn't a case of genetic discrimination: there was no one involved to discriminate against. This was the case that wasn't.

The Burlington Northern case clearly demonstrated that genetic testing follows the loaded-gun theory: if you give enough people loaded guns without clear instructions on when and how they should be used, it is inevitable that at some point someone, somewhere, is going to shoot somebody. By not educating health professionals about genetic testing, and about how minuscule the odds are of this rare genetic disorder being a culprit in carpal tunnel syndrome claims, the safety lock was off the trigger.

A small number of other genetic discrimination cases have come to light since the Burlington Northern incident. One was the case of Heidi Williams, a Kentucky woman, and her two children, Jayme and Jesse, then ages eight and ten, respectively. In 2004, she spoke before Congress about genetic discrimination directed against her family. Heidi has *alpha-1-anti-trypsin deficiency* (AAT). In this disorder, a protein that inhibits enzymes remodeling lung and liver tissue is missing. In the long run, it can cause a patient to always be out of breath, become jaundiced, or die from liver failure. If damage to the lungs or liver threatens to get out of hand, this disorder is treatable by enzyme replacement therapy.

Importantly, in this case, AAT is a recessive disease. This means that although Heidi Williams was affected because she had two genetic mutations in the AAT gene, her children, each of whom carries a functional copy of the gene, are not at risk for AAT and its associated problems.

In the summer of 2003, Heidi was watching television and saw an advertisement from the insurer Humana, Inc., offering affordable health-care insurance. "Call this toll free number," the ad urged, and so she did. The Humana agent asked some questions, computed her statistical probability of making Humana money, and gave her a policy quote of $105 per month for both children. Heidi was interested and told the pleasant young woman on the other end of the phone that she wanted to begin coverage as soon as possible. The agent then asked more questions, including not surprisingly, questions about pre-existing conditions.

"I relayed to the young woman, under a threat of a fine and incarceration for falsifying information, the fact that my children were carriers of the genetic disorder called alpha-1-antitrypsin deficiency, or AAT, a liver deficiency that can progressively affect the lungs, liver, or both, but that my children, unlike their mother, who is lung symptomatic, would never suffer from any aspect of the disorder," Heidi explained in her congressional testimony.

There was a pause. The agent asked her to hold on the line and then had Heidi tell her story to a supervisor. She explained about the genetics, that her husband was not a carrier, that her children would never suffer from AAT, and that they are exceptionally healthy and active children.

Silence. "Please hold the line," the supervisor told her. Heidi then repeated her story to a senior supervisor, and then, transferred back to the agent she had first spoken with, was told she would receive a reply within twenty-four hours. Exactly one week later, Heidi received a letter stating that the insurer Humana was refusing to cover her children due to their AAT status, and for no other reason. This was the second time her children had been rejected for health insurance.

Heidi told her story online to the Alpha-1 Lungs and Life Chat Group, a social network for AAT patients and carriers. Her experience was conveyed to the Genetic Alliance, a larger nonprofit group focused on helping individuals with genetic diseases. The organization approached Heidi about publicly coming forward with her story. Heidi agreed. She was connected to an AAT specialist physician and a prominent Washington, DC, law firm. Together, in August 2003, they wrote an appeal to Humana. Months passed, with Heidi's daughters still without any health insurance. In February 2004, Humana replied to the written appeal and confirmed that her children were being rejected for health insurance coverage "only on the basis of their AAT carrier status and nothing more."

Humana eventually did offer both of the Williams children health insurance, but only after the *Washington Post* picked up the story and started interviewing Humana employees about the case. Needless to say, Heidi's children are now covered by a different insurer. "Humana, Inc., made me feel guilty and ashamed for needing to know my children's genetic status," she said in her 2004 congressional testimony. "I should not have had to spend the better part of six months wondering if the decision to have my children's genetic status verified by their pediatrician was a huge mistake." (A Humana spokesman told CNN in 2007 that "it was a misunderstanding, not discrimination. It was a paperwork error that was later discovered and corrected.")[23]

Because of situations like Heidi's, in May 2008, President George W. Bush signed into law the Genetic Information Nondiscrimination Act, or GINA, which had been stuck in the congressional lobbying maze for more than twelve years. The bill passed the Senate 95–0 and the House of Representatives, 414–1. The only nay vote was from Congressman Ron Paul.[24]

GINA prohibits genetic discrimination in health coverage and employment. Health insurers and employers are not allowed to request genetic information to be used in any of their decisions, and parents don't have to purchase health insurance before their children are conceived in order to avoid later rejection because of a "pre-existing disease" clause. The Genetic Alliance called this landmark in the history of genetics in the United States "the first civil rights bill of the new century."[25] But, GINA does not protect against discrimination involving life, disability, and long-term-care insurance, and it does not apply to companies that employ fewer than fifteen employees.

However, in some states there are additional laws that provide further protections beyond GINA. The Council for Responsible Genetics is a good source of information about the specific protections provided in each state.[26] For example, Cal-GINA in the state of California provides certain protections against discrimination in life, long-term-care, and disability insurance as well as in health insurance and employment. A study of cancer genetic health professionals found that before GINA, 26 percent stated they would perform genetic testing using an alias because of concerns over potential discrimination, whereas after GINA's passage, only 3.2 percent said they would consider using an alias.[27]

Since the bill was signed into law, there have been more than seven hundred allegations of GINA violations. Many of the cases show the ambiguity in US civil court's requirement for a preponderance of evidence proving bona fide genetic discrimination.

Plaintiff Pamela Fink, a public relations director for a Stamford, Connecticut, company called MXenergy, had two sisters, both of whom developed breast cancer. In 2004, Fink had a genetic test showing she carried a mutation in BRCA2 (which causes increased risk of breast and ovarian cancer). In 2009, she decided to have a prophylactic mastectomy to reduce her breast cancer risk, and spent four weeks away from her job while she recovered.[28]

During her recovery period, MXenergy hired a family friend of the CEO, whom he personally recommended. When Fink returned to work, the CEO's friend stayed on staff and, in fact, became Fink's supervisor, a position that hadn't previously existed. Fink felt she was in danger of being losing her job and filed a complaint with the company's human resources department, which called her complaint unsubstantiated. While Fink was preparing to take off ten days for another medical procedure, she received a negative performance review. After she returned to work, she was terminated. When Fink sued her employer for what she alleged was a GINA violation, MXenergy denied any discrimination charges, though without specifying the reason for termination. The case was brought to the attention of the US Equal Employment and Opportunity Commission, which didn't pursue the case.

Fink's situation illustrates some of the ambiguity involving genetic discrimination. Was she fired because she carried a BRCA2 mutation, and this small company didn't want to have their medical insurance rates raised? Was she fired because the CEO wanted to do a family friend a favor? Was her replacement a better employee? Again, the gray areas obscure the true picture, and we will likely never know definitively what happened.

As Robert Klitzman, a bioethicist at Columbia University in New York, wrote in the *Journal of Genetic Counseling*, "Discrimination can be implicit, indirect and subtle, rather than explicit, direct and overt; and be hard to prove. Patients may be treated 'differently' and unfairly, raising questions of how to define 'discrimination.'"[29]

This case reminded me of one professional colleague who had been providing genetic counseling to cancer patients for more than a decade. She didn't want me to use her name for reasons that will become clear later in this chapter. Let's call her Tanya. Tanya said she could recall a number of cases in which cancer patients who carried genetic mutations told her they were fired from their jobs or lost their health insurance after being diagnosed with cancer, but none had raised discrimination concerns before their cancers were found.

I have been a clinical geneticist for almost twenty years, stretching back even before the completion of the Human Genome Project, and I have practiced on both the East and West Coasts. None of my patients has ever told me that he or she was a victim of overt and explicit genetic discrimination (if someone had, I would have assisted in filing a GINA suit). Based on this anecdotal evidence from my experience and my colleague's, such claims of discrimination appear to be extremely uncommon, and unfortunately, there are no definitive studies in the scholarly literature to inform us. However, implicit, more subtle discrimination is another story. I have heard these concerns about workplace discrimination voiced by patients who carry genetic mutations, particularly in what can be extremely competitive and tough professional job markets.

In fact, fears of genetic discrimination are commonplace, expressed by a majority of patients who have gene testing. Studies in 2005 and 2014 suggested that about 40 percent of individuals might avoid genetic testing specifically for this reason.[30] This is the most common conversation I have with my patients, part of routine informed consent for any genetic testing before it is performed.

We will have a hard time getting robust data on how commonplace covert genetic discrimination is. It is perhaps telling that my genetic professional colleague "Tanya" did not want her real name used. She fears her employer, a major medical center, or perhaps even her patients will be angered by her views and damage her career. She fears workplace discrimination will result if she expresses her views on a different type of workplace discrimination. Thus, if people believe discrimination is common for many different nongenetic reasons, then genetic discrimination is a natural (albeit novel) extension of an ongoing problem, the fear of which can never fully be eliminated.

While GINA does not explicitly protect US military personnel, the Department of Defense in 2008 subsequently enacted largely similar policies. However, before time there were an estimated 250 discharges each year due to identified genetic diseases.

For example, Erik Miller was an Army Ranger for seven years and was deployed in Afghanistan. During his tour, he developed back pain.

This led to a diagnosis of von Hippel Lindau (VHL) syndrome, estimated to occur in about three out of every one hundred thousand people. The VHL gene is involved in sensing oxygen levels. In VHL, faulty oxygen sensing causes excessive blood-vessel growth (an attempt by the body to increase oxygen) and tumors, among other problems. Miller had multiple tumors that were compressing his brain and spine. Because his disease etiology was genetic, the army classified Miller as having a "pre-existing condition" and discharged him in 2005, leaving him without health insurance or disability benefits. However, he later appealed this decision and had his discharge benefits reinstated.

Given all the potential as a powerful force for reducing suffering from the many thousand genetic diseases, how are we going to balance genomic progress with the potential for creating new ways for people to discriminate against their neighbors? A few years ago, I was on an FDA panel that discussed the pros and cons of direct-to-consumer genetic testing (which can be bought over the Internet and does not require a doctor or genetic counselor to serve as an intermediary). FDA staffers, academics, and genetic-testing company scientists offered several thoughtful scholarly presentations. One session allowed the public to participate. Anyone could come and give his or her two cents' worth on the topic. One man from a law and genetics public interest research group stood up and chastised the panel. He stated that we were asleep at the wheel, endangering the public, unconcerned with helping people. Then another participant, a former FDA staffer who now worked for a direct-to-consumer genetics company, told us we were over-paternalistic, strangling the industry with regulation, preventing people from obtaining life-changing information, and so on. We were volunteering our time to have the pleasure of being abused from both ideological sides of the debate. However, in the end we have to include everyone at the table in order to develop more comprehensive federal policies so that patients can be protected from genetic discrimination and legitimate testing activities can continue to be used and to save American lives.

The point here is that because of its declining costs and its increasingly mainstream status in medical care, the number of people who could benefit from whole genome sequencing has the potential to soar.

Yet, our laws still do not protect the American public from genetic discrimination as widely and effectively as they should.

Before whole genome sequencing becomes widespread, we urgently need stronger federal laws protecting individuals from genetic discrimination in life and long-term disability insurance, something the United Kingdom has been looking at since the 1990s.

The American, European, and Scandinavian eugenics laws and state-sponsored forced sterilization occurred only a few decades ago, in the 1970s and 1980s. We are not in a position to say, "It cannot happen here," because the history of eugenics programs in the United States, Canada, and Europe shows that dreadful ideas can sometimes achieve mainstream acceptance. While it seems highly improbable, indeed almost inconceivable, that genetic discrimination could manifest itself today in developed countries with extremist policies such as new programs for forced sterilization of genetic "undesirables," it is not completely out of the question that genetic discrimination could arise in the future during times of societal crises in a more subtle form. Historical examples of genetic discrimination show that it often arises because of confusion, misinformation, or financial incentives, rather than from overt prejudice. When we consider widespread whole genome sequencing for American babies, history suggests that society can look very different forty years down the road, in ways that we cannot anticipate, and that may have very serious negative implications for babies whose genomes show "defects."

While it is difficult to imagine the emergence of a genetically defined underclass discriminated against in employment, government social program benefits, insurance, or other benefits, more subtle types of genetic discrimination could perhaps resemble the subtler forms of racism, gender bias, religious intolerance, and homophobia in our societies.

NINE

The Tilted Driver's Test

Learning to drive is a rite of passage for teenagers and their families across the United States.

To date, the playing field for teens learning to drive and passing the all- important licensing road test has been pretty much a level one. Coming from an affluent rather than middle-class family wasn't much help in being able to, say, proficiently parallel park the car during the road test, or know which lanes to use when navigating between multi-lane streets.

But companies ranging from the technology giant Google to some luxury auto manufacturers have already begun introducing an array of new sensors and types of artificial intelligence gadgets that allow cars to do everything from parallel parking themselves, to warning drivers of pedestrians or other vehicles in a blind spot, to braking automatically when a sudden hazard appears ahead. At least one German luxury brand has an optional system now available that allows the car to literally drive itself, stop-and-go style, in the kinds of low-speed traffic congestion many commuters face routinely. And Google has also been testing driverless cars that can operate all the way up to highway speeds in California.

It's probably not likely that any state is going to let a teenager who happens to have access to such a wonder-vehicle breeze through a driver's test by letting the car itself handle all the toughest parts robotically. But imagine the reaction from less well-off kids if a handful of the affluent, whose parents could afford such wonder-cars, drove up in them for their driver's tests. Some of the kids driving middle-class cars without technology packages might respond with the leitmotif grumble known to all parents: "It's not fair." In this instance, the kids would be right.

We live in an age of wealth disparity, with the wealthiest segment of the population owning an ever-growing, oversized slice of the world's riches and financial assets. Recent advances in genetics and the incorporation of genomic knowledge into patient management will have powerful effects, whether positive or negative, on the "fairness gap" between the genetic haves and have-nots across the planet, and this will likely reverberate across generations. The inequalities of American life, including in our health-care delivery system, have been with us for a long time. What is striking about emerging treatments for genetic diseases is how dramatic the inequalities could turn out to be, with the potential for some patients to be left with little access to relevant care, others to have limited access because they are fortunate enough to have the right insurance, and a third group, the very wealthy, having complete access: the medical corollary of "the haves, have-nots, and have yachts."

There are now more than a thousand conditions for which genetic tests are widely available, with yearly increases of about a 10 percent in the number of new genetic tests and a 20 percent increase in gene-based diagnostic tests, compared with about a 2 percent annual increase for nongenetic medical tests. The ever-increasing volume in genetic testing has significant implications for a reshaping of American society in the future.

The cost of genetic testing in the United States typically runs at present from less than one thousand dollars to several thousand. These tests are all too often out of reach for patients who lack "the right kind" of medical insurance coverage, even when their histories present a strong case for genetic testing. Depending on the specifics of the individual patient, the insurance plan, and even the state of residence, appeals to insurers sometimes win, but often only after patients or their doctors spend a tremendous amount of time haggling, letter writing, and outright begging.

The prices of these tests are clearly too high and need to come down. Yet, even after skirmishes between patients and their medical teams with insurers or Medicaid or Medicare are won, the chronic war for long-term medical management and treatment still looms. We are facing a world in which the prices of treating chronic genetic conditions that can require therapies for life dwarf the cost of testing. For example, the annual cost of medicines to treat disorders such as Gaucher disease (described in chapter 1) can easily top $400,000 per annum.

In such cases, both the uninsured and middle-class insured find themselves tossed in the same bin together, unable to either pay for or get insurance coverage for their ongoing treatments, while the very wealthy can simply choose to pay for follow-up and treatment out of pocket. To return to the driver's test analogy, it's as if the wealthy and middle class haven't been so concerned about kids whose families can't afford cars. But now insured members of the middle class find themselves largely bifurcated from their wealthier cousins, in the back seat with their less well-off brethren and likewise mumbling to the front seat that "it's not fair."

Genetic diseases can have dramatic effects on a family's long-term prosperity. The costs of chronic-disease care can obviously exhaust even an extended family's resources. Taking care of children who require intensive therapies can mean the difference between whether one or both parents can work outside the homes. For genetic disorders that affect adults and not kids, the families of people carrying mutations clearly may suffer financially if a parent is unhealthy. The consequences of chronic disease, long-term disability, or premature death can similarly have an impact extending across generations.

Timothy Powers was a young man I saw during my clinical genetics training around the year 2000. A kind, tall, lanky fellow with blond hair, brown eyes, and a sharp wit who was a student at an elite private university in the Washington, DC, area, he had already had three heart attacks by the age twenty. His father had died of a heart attack when Tim was a toddler, a terrible event that had a major impact on Tim's psyche.

Tim had a severe form of familial hypercholesterolemia. In this disease, the body has a mutation in a gene that makes a protein that in turn helps dispose of cholesterol, which otherwise would build up in the circulatory system and elsewhere. Tim had small bumps on his elbows and his eyelids that were deposits of this fatty substance. The same cholesterol also caused similar bumps in the cells linking blood vessels to his heart, leading eventually to a blockage that had caused part of the heart muscle to choke and die. The less severe form of familial hypercholesterolemia, in which men can develop heart attacks in their thirties and forties, is as common and occurs in one in every five hundred people, while the more severe form that Tim had occurs in about one person per million.

Tim was being treated with diet, drugs, and exercise to lower his cholesterol. His treatment included regular sessions during which he was hooked up to an IV to have the cholesterol filtered from his blood. He was not a candidate for a liver transplant, which would have been able to reduce his cholesterol but would have required lifelong immunosuppressive drugs that would make him susceptible to infections, brittle bones, diabetes, and potentially, cancers of the blood and skin. Tragically, this accomplished young man with a bright future would die from his disease about a year after I saw him.

If Tim had been found at an early age to carry a familial hypercholesterolemia mutation, he might have been started on cholesterol-lowering therapies at a younger age and lived into his thirties. Today we have new powerful, scientifically elegant, robust drugs approved specifically for familial hypercholesterolemia, making familial hypercholesterolemia is a highly actionable disease. If Tim were still alive today, I often wonder whether he might have been able to live into middle age or perhaps longer.

Several years ago, a study estimated that familial hypercholesterolemia genetic testing to screen all sixteen-year-olds could eliminate some deaths by the age of twenty-six for about $700,000 per death averted, including ten years of statin medication starting at age sixteen.[1] This analysis, from the United Kingdom, budgeted about $400 per year for statin drugs, a price that is now lower with generic drugs available in both the United Kingdom and United States. However, the two newer drugs that are effective and approved specifically for treating this disease ring in at $150,000 to $300,000 per year, *several hundred times more.* If used for all familial hypercholesterolemia patients, this would likely cost tens of millions per death averted. The cost could further spiral out of control if we wanted to provide the best therapies for the increased number of patients who would be diagnosed with universal screening.

The newer drugs for treating familial hypercholesterolemia are much more effective than the older, cheaper ones, but the price is staggering. With costs so high, tough choices would likely have to be made, meaning that some kids might be treated with inferior drugs or perhaps not be treated at all.

In a world of limited resources, cost-effectiveness analysis is commonly used to measure how much of a bang a patient gets for a given medical buck. For both nongenetic and genetic disorders, health economists typically use a so-called *quality-adjusted-life-year* (QALY) ranking to evaluate the benefits of different therapies, all of which compete for same finite number of dollars. Each year in perfect health is given a value of 1.0, and being dead a value of zero. QALYs can also be less than zero for certain conditions such as being unconscious in a coma and on a ventilator. Accepted ranges for QALY spending in the United States range from about $50,000 to $200,000 to treat a bad disease. For example, the cost of kidney dialysis is currently about $75,000 to $100,000 annually, and the quality of a life for a year on kidney dialysis is ballparked at approximately one half that of a "healthy" life. This computes to a QALY-adjusted cost of spending $150,000 to $200,000 per year for kidney dialysis, which is considered acceptable to Medicare and many private insurers.

Of course, it is difficult, if not impossible, to reach consensus about how to define perfect health. For example, some argue that mental health benefits are undervalued relative to physical pain or disability, while others disagree. The impact on overall quality of life for family and other caregivers also is not included in the calculation.

When it comes to determining a survival benefit for different therapies—how long someone lives with or without a given treatment plan—we can be extremely precise. Years, months, weeks, days, and even hours, minutes, or milliseconds are things we can measure with precision and consensus. Yet, quality of life is different. Get ten people in a room and ask how they define quality of life, and you may get eleven different answers. This can make assigning QALY values very challenging for health-care economists and policymakers who decide which procedures to cover and which not to. This collection of disparate viewpoints means opinions are largely the weights in coverage debates balancing the merits of coverage for different therapies improving quality of life but not necessarily length of life.

It was a typically perfect, sunny day in Southern California when I met Sean Rodriquez in a clinical genetics clinic there. Sitting in front of a wall-length salt-water aquarium in the clinic waiting room, Sean looked very out of place, literally a fish out of water. He was a real-deal California dude: muscular, in his early twenties, ruggedly handsome, bronzed with natural sun and ocean highlights in his hair. He was wearing a Bailey pork pie hat, a University of Southern California Trojans T-shirt, cargo shorts and had headphones in his ears. He definitely had a certain look going for him, one I saw all over the local beaches and malls, like a Jason Mraz/Patrick Swayze heterozygote. His appearance was quite different from most of my patients. Why was he here to see me? It crossed my mind that maybe he had accidentally come to the wrong clinic.

It turned out that Sean had a mild form of *hereditary angioedema* (HAE), a rare genetic disorder in which a gene that helps control inflammation and swelling (like when you stub your toe in the doorjamb) is defective. In patients with this disorder, inflammation can get out of control, whether it's caused by pressure on a body part (striking that doorjamb) or an infection, or has no clear cause. The most worrisome site of this inflammation is the throat, a potentially life-threatening emergency since someone with HAE can suffocate to death as the airway constricts. This could require insertion of a breathing tube or even a tracheotomy. The swelling can also occur in other soft tissues, such as the intestines, the soles of the feet, the buttocks, and other soft, fleshy places.

Sean worked in a surf, skate, and ski equipment shop in Orange, California. He was passionate about all three sports. While still in a suburban Los Angeles high school, he had placed well in some skating competitions. Trauma from skating mishaps can certainly cause HAE attacks, but he was careful, skilled, and just couldn't give the sport up, despite loud protestations from his family and doctor. He had a few emergency room visits after some falls, but nothing that had really gotten him into trouble. After high school, Sean went to a local community college but had taken a leave after a few quarters. Working in a skate shop was a dream job for Sean—except that it left him with no health insurance.

One of the lessons we learn from practicing medicine is that each patient's particular circumstances must be taken into account. Sean had heard from HAE support groups about powerful new medicines called *C1-inhibitors* to prevent angioedema. However, since these drugs were still under patent protection, they cost over $100,000 per year. As far as Sean was concerned, these drugs might as well have been approved for use on Mars instead of Earth.

Contrary to my expectations, when Sean first revealed that he had the HAE disorder, he had not yet had any life-threatening problems such as sudden swelling in his throat. However, this otherwise robust young man was having certain lifestyle issues. In his own words, he was hanging with his friends, living "the life" in the Orange County beach communities' singles scene, and, he explained, pretty successful in the "hookup" department. But he was having problems with his "privates," in the form of extreme pain from his HAE. He couldn't exactly predict when it would or would not happen, but sometimes he'd end up with painful swelling of his genitals after intercourse. The pain was both physically and psychologically interfering with his sex life, leaving him with performance anxiety even when HAE wasn't bothering him. A relative who had had a similar problem had ended up joining the priesthood (perhaps seeing this as a personal message from the Almighty rather than a genetic issue), presumably before anyone in the family had heard of HAE.

Sean wanted a prescription for C1 concentrates to use prophylactically before his hookups. From one perspective, Sean just wanted to live a normal Southern California young adult life like all his friends. Unfortunately, his only health coverage came from California's version of Medicaid, called MediCal, and MediCal had a different perspective: it wasn't willing to pay for the expensive C1-inhibitors.

Sean was left like a starving Dickensian orphan staring into the window of a well-stocked bakery in the morning. There these effective, new twenty-first-century medicines were, just behind the glass, out of his reach. From his viewpoint, this was the worst thing that could possibly happen to him, and he felt that his quality of life suffered greatly.

I was able to start Sean on a much cheaper therapy called Danazol to help his symptoms. Danazol is a steroid that had been used to help reduce HAE attacks for many years, but it caused problematic side effects. Sean had a reputation for being very relaxed, but after starting the drug, he suffered from serious reactions, including episodes of rage that got him in trouble at work with both managers and customers, and also risked getting him into bar brawls. Getting punched in the face in a brawl was something potentially very dangerous for persons with HAE. Although he was otherwise a social success, given his looks and pleasant personality, Sean became depressed as a result of having a genetic disease but no means to obtain the best treatment.

As these cases and countless others illustrate, American medicine often is much more effective at innovation than at long-term care for chronically ill patients. Just as turning the spigot on a firehose means that someone is going to get wet, the increased amount of genetic testing inevitably means more people will be diagnosed with diseases like familial hypercholesterolemia, HAE, and many other actionable DNA-based diseases that previously went largely undiagnosed. Under the current health-care system in the United States, the costs of chronic therapies could be staggering.

Another expensive genetic technology that has become more widely adopted, and which will reverberate into future generations, is third-party reproduction—that is, various techniques for producing a child that involve more people than a conventional father-and-mother model, involving surrogates and, usually, *in vitro* fertilization (which is described in more detail in chapter 3, "Altitude Sickness").

The third party can be a man other than the father-to-be who provides sperm, or a woman in addition to the mother-to-be who donates her ova, the mitochondria from her ova, her uterus for the actual mother's ova, or some combination of the above. *Surrogacy* refers to a woman who carries a child genetically unrelated to her (*gestational surrogacy*) or donates her ova (*biological surrogacy*).

In theory, a newborn could have up to *six* distinct parents. The first is a genetic father who provides the genetic material (in the form of sperm) to create the child. The second could be a social father who helps nurture and raise a child even though they might be biologically unrelated. Next are up to four mothers. The first provides the genetic egg (or ovum). The second provides the mitochondria DNA for the ova. The third is the gestational mother who carries the child, and the fourth is the social mother who helps raise the child. Social parents, of course, could also be of the same gender or trans-gender. Divorce and remarriage could further increase the social parent tally as the child grows up.

While I have a hard time imagining that a sexta-parent child would be attempted (in particular, in wouldn't make sense for social parents to choose a donor whose eggs needed mitochondria from another donor), my experience tells me to expect the unexpected and that some version of this multi-parental arrangement is likely to occur somewhere at some point in time. (In fact, I speculate that over the next couple of decades, a human clone will come into being, largely through the use of existing technologies and human stem cells.)

In conventional surrogacy, a female surrogate is inseminated with sperm of the male partner of a couple (or one of the males from a same-sex couple), or from another sperm donor. The surrogate then carries the baby to term.

In gestational surrogacy, a female surrogate has an embryo implanted in her uterus, where the embryo was conceived in an ovum provided by another woman (such as a couple's female partner, one of the females in a same-sex couple, or a single mother). The embryo implanted in the uterus of the surrogate mother can be fertilized with sperm from the male partner, one of the males from a same-sex couple, a single father, or a sperm donor. The first child born from ova donation was reported in Australia in 1983. Since then, almost fifty thousand children resulting from donor ova embryo transfer have been born in the United States, involving, as parents, many well-known figures such as the popular songwriter and musician Sir Elton John, the actress Nicole Kidman, and actor Neil Patrick Harris. About one in six births using in vitro fertilization employ donor ova.

Across the globe, there is great variation among different nations regarding the social and legal acceptability of gestational surrogacy. In Germany and Italy, gestational surrogacy is banned completely. In

Canada and France, it is legal only if performed without financial compensation for the surrogate, for example, if the gestational surrogate is a relative of one of the biological parents. In Spain, the procedure is legal and does allow payment for expenses but not additional compensation. In most US states, egg donors may be compensated. However, paid egg donations are illegal in New York State, as well as in Arizona, Indiana, Michigan, Nebraska, and the District of Columbia. In some countries, such as Spain, egg donors are disproportionally recruited from among women in poverty or near poverty. Surrogacy can also affect nationality and citizenship. In the United States, the State Department has a policy that "in order to transmit US citizenship to the child born abroad . . . there must be a biological relationship between the child and a US citizen parent."[2] Thus, for a child to be an American citizen, either the biological father or mother must be a US citizen, regardless of who is the gestational surrogate.

Compensation for donor ova in the United States can be substantial. Scanning different newspapers from elite colleges with low acceptance rates, one finds advertisements offering $25,000 or more for "right stuff" eggs from young women, with compensation varying according to personal characteristics, talents, and ethnic background. Requirements (in addition to pristine personal and family medical histories) almost always include height above five feet ten (which is taller than I am). After height, other requested characteristics of egg donors on donor applications include admittance to elite colleges, specific hair color and texture, specific eye color, specific skin tone and color, high IQ test and SAT math scores, documented varsity athletic prowess, musical accomplishments documented by recordings, and career choice (extra points for being pre-med). Some of the donor application forms ask for more nuanced information, such as personality type (outgoing, studious, etc.). Some of these latter, softer characteristics are, of course, dependent not only on genetics but also on the environmental influences in a person's life (e.g., whether the donor's parents supported musical training at an early age; whether she was raised with siblings) and, so, may be problematic to rank. Although computational simulations of desired children's physical appearances are available for sperm bank matches, I could not find anyone who would admit to using this type of software for egg donors.

Like brokers for high-end real estate, private planes, and yachts, there are egg brokers in New York and most other large American cities. A prospective couple can meet in person or videoconference with egg donor candidates, their current children, and other family members. Because the United States has the highest market prices for donor eggs, it also has the largest pool of potential candidates. Popular egg donor ethnic groups in high demand reflect market forces. In addition to Caucasians, the highly sought after donors include Japanese, Koreans, Indians, Han Chinese, and Ashkenazi Jews.

Recently, mitochondrial surrogacy has also become possible. Mitochondria are organelles inside cells that use oxygen to provide adenosine triphosphate, or ATP, which is the universal currency of energy for cells to perform their daily activities. Mitochondria as organelles are evolutionarily descended from symbiotic bacteria that invaded and co-opted ancestral single cells about a billion years ago, before they evolutionarily diverged into plants and animals. Because mitochondria are derived from bacteria, they have their own genome that encodes thirty-seven genes and is about one hundred and eighty thousand times smaller than the human genome. The mutation rates for mitochondrial genes are about ten- to twenty-fold higher than those of human genes, likely because of damage from oxygen (as described in more detail in chapter 3, "Altitude Sickness"). In fact, increased mitochondrial damage that occurs with aging is thought to be a major cause of age-related female infertility and genetic diseases caused by faulty chromosome segregation (whose molecular motors are high octane and require significant amounts of cell ATP energy), such as Down syndrome.

To increase fertility rates in ova from reproductively mature women, cytoplasm from eggs donated by younger women can be transplanted (a procedure referred to as *oocyte augmentation*, or sometimes *ooplasm transplant*). Both donor and recipient mitochondria can be detected in children born from oocyte augmentation pregnancies, reflecting the genetic contributions by two mothers to the same child.

In studies of mice, rabbits, pigs, and cows—as well as in smaller, non-definitive preliminary studies in people—ooplasm transplant was associated with increased successful pregnancy rates and live births. However, there are still many questions regarding the long-term safety and ethics of oocyte augmentation. While this procedure is currently available in several countries, it is not yet approved for use in the United States. In the United Kingdom, oocyte augmentation pregnancies are being performed for women who carry mutations that cause mitochondrial genetic diseases.

Economically, the situation is nuanced and complex. Consider that on one hand, these technologies could reduce the burden of genetic diseases and associated costs, even while, on the other hand, the sheer costs of expensive medicines to treat genetic diseases has the potential to balloon out of control. In summary, the ethics of these techniques to reduce the chances of having genetic diseases in embryos continues to be debated and likely will be for a long time going forward.

To delay or, better yet, avoid a looming crisis in medical spending from the increase in genetic testing, we need three things. First, we need to invest in new breakthrough technologies that can treat genetic diseases more cheaply. This will take patience, but it would be the best solution to this problem. Second, in the meantime, the price of current therapies for genetic diseases must be negotiated down to sustainable levels, especially as the volume of patients and related drug sales increase. Third, we must focus more on how to manage chronic diseases, including both genetic and nongenetic disorders, more effectively.

Failure to address these problems has the potential to transform the United States into a barbell society with, on one end, the haves, who can afford high-priced effective therapies for genetic diseases and, on the other, the have-nots, with a wide distance between them. In the end, it is important that we use genetic knowledge to help level the playing field and increase the diversity of the American community, not tilt the incline further. The more genetic diversity we have, the greater the probability of achieving hybrid vigor. In *hybrid vigor*, genetic traits are enhanced because of mixing the genetic contributions of parents so that the offspring can exceed both its parents and even produce traits that were not present in either parent. By properly treating patients who have genetic diseases and allowing them to prosper, as a collective society America can create such variety that whatever challenges come before us, we will can draw upon the American ethnic and genetic diversity that is one of our country's greatest strengths.

TEN

Fingerprints, Written in Blood

For Kirk Bloodsworth, an ex-Marine and commercial fisherman, the nightmare began one day in July 1984 in the wee hours of the morning. At 2:15 a.m., he recalls, "there was a big bang on my door. Boom, boom, boom. I open the door and there's a flashlight stuck in my face. I step outside.[1]

"A voice behind the flashlight's glare said, 'Mr. Bloodsworth, you're under arrest for first-degree murder, you son of a bitch.'

"They read me my rights, and that was the last time I saw my small town for eight years, ten months, and nineteen days."

Kirk Bloodsworth had been accused of—and subsequently was tried, convicted, and sentenced to death for—a brutal crime that, as it turned out, he had never committed. In the end he would turn out to be the first American initially sentenced to death and later exonerated based on DNA evidence.

Thanks to the entertainment media, particularly television programs such as *CSI*, criminal forensic investigation may be the most familiar use of DNA analysis for most Americans. But just as the stories of different people in this book have illustrated that using DNA in medical settings is both an enormously valuable tool and a process fraught with complexities, so too is DNA analysis in the science of forensics.

DNA has allowed the criminal justice system to capture and punish heinous wrongdoers. It also has exonerated a surprising number of innocent but nevertheless convicted and imprisoned victims. But just as the ability to perform a difficult surgical procedure on one patient doesn't mean it can be extrapolated to work on all patients, gaps, omissions, problems, and misperceptions in the granular details of forensic science can make for real problems as well.

We see the world not necessarily as it is, but as we are—that's a mantra of our colleagues the cognitive brain scientists. The human brain is hard-wired to see patterns in randomly collected data, even when there are none, a genetic predilection of intelligent animals whose technical term is *apophenia*. Examples include phallic objects in Rorschach tests, ghosts in nighttime shadows, and unidentified bright objects on smudgy MRI scans.

Evidence is *circumstantial* when it requires inference and at least a bit of speculation to glue it to fact. By its own definition, circumstantial evidence can be explained by several different potential scenarios. An eyewitness sees an object that may have been a weapon, but it might also have been an aluminum pipe, a telescope, a reflection from a window in the same line of sight, or nothing at all.

Today, the ability to locate, secure, and vastly amplify even almost infinitesimal samples of DNA has advanced to the point where a few microliter drops of chewed tobacco spit or sweat, or some flakes of dandruff can place a criminal at a scene walled off by yellow tape and marked by chalk. And it's become almost expected—by the media, but more importantly by courthouse juries made up of people exposed to news and television crime drama—that investigators produce DNA evidence as proof of guilt, particularly for some violent crimes like sexual assault and murder.

The technological ability to use DNA "fingerprinting" as evidence is, surprisingly, a relatively recent event. In fact, about a third of a century passed between Watson and Crick's description of DNA in the 1950s and the discovery and subsequent use of forensic DNA identification in a criminal case.

The initial discovery in 1984 that DNA could be used this way to identify potential persons of interest (and unintentionally, their relatives as well) came as a surprise even to the people who made the discovery—a British geneticist named Alec Jeffreys and his research team at the University of Leicester.

"The last thing on our minds was forensic identification," Jeffreys said in a 2013 interview with the journal *Investigative Genetics*.[2] Jeffreys said the discovery was driven by "academic curiosity." He was particularly interested in seeing if he and his team could find "genetic markers far more informative" than anything that had been seen before for human beings.

The unanticipated breakthrough came at about nine o'clock one September morning. Jeffreys had just developed an X-ray film showing pictures of radioactively labeled DNA fragments collected from different members of several families run side by side. Lab workers had taken blood and biochemically purified the genomic DNA in a series of chemical extractions and precipitations. Then Jeffreys's lab used special reagents called *restriction enzymes*, which had been purified from bacteria and which only recognized specific sequences, to cut the strands of DNA. Bacteria have evolved that make these restriction enzymes to cut the DNA of invading viruses, called *bacteriophages*, that have specific defined sequence motifs. But, as Jeffreys knew, remarkably, the same enzymes also can recognize and specifically cut human DNA with the same, often palindromic motifs, such as the sequence GAATTC.

Jeffreys was slicing out sections of DNA that didn't otherwise seem very useful biologically: genetic sequences that don't actually contain any coded instructions. In general, these expanses of genetic material are seen as "junk," loathed as biological nonsense by the computational biologists who study gene sequences because they massively clutter up analysis of the unique base-pair combinations that actually determine biological traits. The stretches of code, called "satellites," are stuck between the sequences that actually contain instructions, like deserts between oases, monotonous, repetitive, stuttering stretches that must be endured during genome scans. For many kinds of studies, these elements are masked out and excised as no more than an afterthought, like having to take out the garbage after a family dinner.

But an old adage tells us that one person's junk can be another's treasure. Well represented among junk DNA chromosomal refuse are literally millions of short, elements of the non-coding DNA: those satellites, and their junior relatives, mini- or micro-satellites. *Minisatellites* have longer repetitive sequences, and *microsatellites* have shorter repeats, but both can stretch for hundreds or even thousands of bases along a strand of DNA.

During the genomic copying process, when germ cell sperm or eggs are made, various slipping and sliding stuttering errors get introduced. Junk DNA repeats can expand or contract. The same is true for random errors occurring even in non-repetitive elements that aren't in positions in the genome (such as important proteins) critical for health. For each child, there are typically several of these stuttering microsatellites or satellites that differ from those of its parents. Because most of these errors don't influence a person's medical propensity to disease or dramatically affect basic features such as height or intelligence, and thus don't reduce an individual's odds of surviving and reproducing, they remain and can be passed down to future generations. The bottom line: each of us share many of these non-coding satellite sequences with our parents and other relatives; but every one of us also carries some satellite sequences that are utterly unique each individual.

Because satellites have different percentages of A, C, G, and T instances their DNA pieces weigh different amounts and therefore and can separated by density. This made them the gold treasure buried among the junk that became useful for Jeffreys's forensic identification and gene mapping.

Jeffreys used electric current to separate the DNA. In this process, called *electrophoresis*, smaller restriction enzyme-cut fragments move to one end of a gel matrix, while the larger fragments remain on the other side.

The Jeffreys lab then used different satellite DNA sequences as probes labeled with radioactive phosphate to scan the blots by size. Each person's DNA was run in a separate side-by-side lane, each radioactive human genomic DNA finding its corresponding sequence, with the radioactively labeled GATC and longer motifs binding to their CTAG mates. On an X-ray film, the radioactivity made black smudgy side-by-side parallel bands where the different fragments had run, making the film look somewhat like a piece of an old-fashioned player piano roll or an old Univac-era computer punch card.

"I think the penny dropped within about a minute of developing that first X-ray film. . . . My entire life changed in the space of about sixty seconds," he said.

What Jeffreys almost instantly realized was that the patterns that emerged on the X-ray film differed uniquely among individuals, with patterns varying least among immediate family members. It amounted to the first evidence that DNA could be used to identify precisely to whom it belonged and who shared it.

As evident as the notion seems today, it was wholly unexpected. Almost instantly, Jeffreys recalls, he could see that his sudden surprise discovery had applications for determining paternity and that the process would also help wildlife biologists studying biological diversity. And, he also realized, it would be a tremendous tool for crime scene forensics.

Jeffreys recalls that he also quickly realized that science knew precious little about how well DNA in, say, a blood sample would even survive. So he spent that very afternoon pricking his fingers repeatedly, leaving blood samples all over his lab to study what happened as blood dried and aged.

That very night, after he excitedly told his wife, Sue, about the new breakthrough, she almost instantly came up with a new twist on how the technology could be used: in immigration disputes in which authorities doubt or dispute claimed family relationships. In fact, resolving such a dispute would, before long, turn out to be the world's first practical use of DNA fingerprinting.

His colleagues, however, weren't so sure. A few weeks after the discovery, he described what he had come up with, in an informal lunchtime talk to colleagues in his own department at the University of Leicester:

> I went through the biology in these bits of DNA and the shared sequence and how you could potentially pick up lots of these mini satellites at the same time and how the patterns are individual-specific. That was all fine. Then I said, and we might be able to use this for example to catch rapists. And I remember these howls of derision . . . half of my colleagues clearly thought Jeffreys has completely lost the plot here. And that just made me even more determined that, you know, this is going to work.[3]

The technique would be linked to fingerprinting later: At a seminar he was giving at another university, Jeffreys mentioned some of the patterns that had been revealed and his hunch about how they could be used. A friend and colleague approached him afterwards and suggested that it was much like fingerprinting, which immediately struck Jeffreys as an ideal way to explain the technology.

"And that actually was a very wise thing to do," he's said. "If we had called these, say, individually discriminating southern blot mini satellite hybridization profiles, I mean we would have killed the technology absolutely stone dead. It would never have got off the starting block."

It did indeed get off the starting block thanks to the plight of a Ugandan immigrant to the United Kingdom named Christiana Sarbah, who had faced only frustration in trying to convince authorities that her son Andrew was in fact truly her biological child. The dispute had actually begun in 1983, the year before Jeffreys's surprise breakthrough in the lab. Then-thirteen-year-old Andrew had been detained by immigration authorities at London's Heathrow Airport. He had a British passport, but the authorities suspected that some trickery was afoot. His parents had separated, and the lad had spent several months in Uganda with his father. Immigration agents were convinced that either the passport had either been altered or the teenager who landed at the airport wasn't the real passport holder. They eventually allowed him to return to his mother's home in London, but only after a Member of Parliament intervened. Andrew remained subject to deportation.

A British nonprofit, the Hammersmith Law Centre, tried to help with the case, compiling testimonials from relatives, family photographs, and even some biological information. They identified a shared blood group in common to suggest that Cristiana and Andrew were, in fact, closely related. The British Home Office remained skeptical, suggesting that the young teen claiming to be Andrew could be the son of one of his mother's Ghanaian sisters. Mother and son lost an immigration hearing but did win the right to appeal the decision. If they lost on appeal, Andrew, who had grown up and spent his whole life in England with his mother and siblings, would be deported.

After reading a newspaper story about Alec Jeffreys and his recent breakthrough, the Hammersmith Law Centre asked for his help. Using blood DNA from three of Andrew's own siblings (whose parentage no one was disputing), his mother and, of course, Andrew himself, as well as someone not related to the family at all, Jeffreys complied a series of DNA fingerprints.

Like his siblings, Andrew had DNA that shared some twenty-five bands of satellite repeats, a pattern clearly inherited from their mother. Jeffreys was able to calculate the odds that somehow Andrew's real mother was one of the aunts: vanishingly small, less than one in a half million. Jeffreys was also able to reasonably approximate the boy's father's genetic fingerprint by comparing samples from all four of the siblings, looking for bands that the children showed, but Christiana did not. In 1985, the Home Office accepted Jeffreys's conclusion, and in fact announced that henceforth, it wouldn't contest claims of parentage when such compelling DNA evidence could be compiled.[4]

Three years later, Jeffreys's DNA fingerprinting discovery was used for the first time in a murder investigation that eventually solved the rapes and murders of two teenage girls in the Leicester region. Remarkably, it would amount to two firsts: the first time DNA fingerprinting both exonerated one innocent individual and the first time it led to the conviction of the actual perpetrator.

In November 1983, schoolgirl Lynda Mann, fifteen years old, had been sexually assaulted and strangled on a deserted pathway known as the Black Pad, on the grounds of a psychiatric hospital in Narborough. A semen sample revealed that the perpetrator had type A blood. That and the sample's enzyme pattern narrowed the field of suspects to 10 percent of the local male population. But the crime went unsolved.

Then, three years later, a second fifteen-year-old named Dawn Ashworth was also raped and strangled in a nearby village, on yet another remote pathway, this one called Ten Pound Lane. Semen samples revealed the murderer was, once again, a blood type A with the same enzyme profile.

This time the police were able to identify a likely suspect, a mentally disabled seventeen-year-old named Richard Buckland, who worked at the psychiatric hospital. Buckland had been spotted near Dawn Ashworth's murder scene. Brought in for questioning, Buckland eventually confessed to the recent killing, but denied having anything to do with the earlier assault and murder.

Jeffreys, meanwhile, had teamed up with two expert forensic scientists, Peter Gill and David Werrett from the UK's Forensic Science Service (FSS), to develop a technique for extracting DNA from just such crime scenes and then fingerprinting it. They published their first report on the approach in 1985. Gill, the lead author, had by then developed techniques for extracting usable DNA from victims of sexual assault by separating the victim's vaginal cell tissue from sperm cells.[5]

When scientists took a blood sample from Buckland and compared its DNA with the semen samples, it became clear that this disabled young man had not been the perpetrator of *either* crime. Mentally delayed, he may have withered under the hard interrogation of insistent, and perhaps impatient, police officers psychologically bludgeoning him for a confession. Buckland thus became the first person ever to be exonerated based on DNA evidence.

Jeffreys later said, "I have no doubt whatsoever that he would have been found guilty had it not been for DNA evidence. That was a remarkable occurrence."

Meanwhile, baffled police were once again left without a suspect.

What followed, in 1987, was the first mass screening of potential suspects' DNA. Casting a wide net, police attempted to check the relevant genetic code of more than four thousand men in the region who could not come up with alibis for the murders. The investigation managed to get samples from nearly all—98 percent—of the men in the area without alibis between seventeen and thirty-four years old. And once again, investigators came up with nothing.

The breakthrough came that August, when a woman reported to police that she had overheard a coworker bragging that he had tricked the investigators, pretending to be a friend of his in exchange for money when the samples were taken. That friend, a local baker named Colin Pitchfork, was soon arrested and compelled to give a DNA sample. The pattern of the DNA microsatellites in the sample turned out to be perfect matches for the semen samples from both crime scenes. Pitchfork, who had admitted under questioning that he had a compulsion for "flashing" and had exposed himself to hundreds of women and girls, also confessed to the assaults and murders. Pitchfork, the very first criminal in the world identified and convicted thanks in part to DNA evidence, was given a life sentence.

Kirk Noble Bloodsworth, a Maryland man who woke in the dark in July 1984 to the sound of angry police pounding on his door, had been accused of and arrested for the brutal rape and murder of nine-year-old Dawn Hamilton. Today he owes his freedom not only to the DNA fingerprinting technique Jeffreys developed, but also directly to the story of the manhunt for Colin Pitchfork.

In the days immediately after Dawn Hamilton went missing, and was later found dead in a wooded area, police scrambled to find the murderer. Two young boys said they had seen her walking with a tall, thin, blond man with a bushy mustache. A police artist prepared a composite sketch based on the boys' description, and before long received a tip from a woman who said the sketch looked like a guy named Kirk who had recently moved to the area and was working at a local furniture importer. The police obtained a photograph of Bloodsworth and showed it and other photos of men's faces to the boys in a "photo lineup." One boy could identify no one in the photo array as the man he'd seen. The other pointed to Bloodsworth, but said the hair color wasn't right: the man they'd seen was blond, and Bloodsworth's hair was bright red.

That's not all that didn't square with the description. Bloodsworth was far from thin: he carried a thick and hefty 230 pounds on a frame that was six feet one inch. Police called him in for a physical lineup anyway. One of the boys still couldn't identify anyone from the lineup group. The other pointed to a police officer in plain clothes who had joined the group to help fill it up.

But, to police, Bloodsworth seemed to be acting suspiciously. Shortly after Dawn was murdered, Bloodsworth had left town without telling his wife where he was going. She had filed a missing persons report. He would later claim that his marriage was on the rocks, and he just had wanted to get away from his wife. Police found him in the nearby Maryland town of Cambridge where he'd grown up and where the couple had lived until recently.

The next year, Bloodsworth went to trial. By then, the boys had decided that in fact he was the man who'd been walking with Dawn after all, and three other eye witnesses said they'd seen him in the area. Bloodsworth had no credible alibi. And police forensics experts testified that shoe marks on the girl's body matched a pair of Bloodsworth's shoes. In March 1985, he was convicted of kidnapping, sexual assault, and murder, and he was sentenced to death.

Bloodsworth says he spent his early weeks on death row stunned at what had happened to him, frequently weeping. Then he began to read: thousands of books—novels, nonfiction, whatever he could get his hands on. In 1987, he got a new book in the mail from a friend. Called *The Blooding*, by the former Los Angeles police officer and best-selling mystery writer Joseph Wambaugh, the book was about the true story of the hunt for a killer in England. It told the story of how Jeffreys's DNA breakthrough had led to the conviction of Colin Pitchfork.

The true story of *The Blooding* and the scientific basis of DNA fingerprinting gave Bloodsworth new hope. What if there was DNA evidence that could prove that he wasn't the rapist and killer?

Unlocking that evidence would take years. In 1991, Bloodsworth managed to secure a new lawyer, Robert Morin, but Morin at first couldn't offer much hope. He had been told by prosecutors that forensic evidence had been inadvertently destroyed. But Morin was later able to locate critical evidence in the form of the girl's underpants, which had been found hanging from a tree.[6]

(Bloodsworth, by this point, was at least no longer on death row. A few years earlier he had won a new trial after establishing that the prosecution had withheld potential exculpatory evidence. He was reconvicted, but sentenced to life in prison, so the death penalty no longer hung over him.)

By 1992, the attorney had become convinced that new DNA technologies might allow evidence to be unlocked after all.

The *polymerase chain reaction*—a new way to rapidly and vastly amplify samples DNA previously too tiny to analyze—had by then come into play. In 1992, Morin obtained a court's permission to send evidence to a scientist named Ed Blake in California who had pioneered DNA evidence testing in the United States and operated the sole private lab that was testing DNA in crime investigations. (Bloodsworth's family by then had exhausted all of its resources paying for the two trials. Morin paid the $10,000 fee for the analysis out of his own pocket.)

There turned out to be one dried spot of semen about the size of a dime on a pair of panties. With an existing backlog and technology far slower than today's, it took a year for the results to come back. But they were clear; the DNA wasn't from Bloodsworth. Part of the deal Morin had cut with the state in order to obtain the evidence mandated double-checking the findings with the FBI's crime lab. Within weeks, the lab confirmed that the DNA and Bloodsworth's were not a match.

Bloodsworth was released from prison in April 1993 and, eventually, awarded $300,000 to compensate him for the wages lost during the nine years he spent imprisoned, with both the release and the payment contingent on his agreeing never to sue the state for further restitution. He was a free man, but about ten more years would roll by with the crime unsolved. In fact, in 2000, seven years after his release, the two attorneys who had prosecuted Bloodsworth seemed to suggest that some sort of cloud still hung over him. Former prosecutor Robert Lazzaro suggested to the reporters that, perhaps the semen had accidently gotten transferred to the girl's panties from her father's underwear in the family laundry, and that for some reason she'd put on underwear that never did get washed. Or that Bloodsworth was perhaps part of a two-man team that had attacked the girl, with the unknown other perpetrator's semen possibly exonerating a co-conspiring Bloodsworth.

In early September, 2003, Lazzaro's co-prosecutor in both trials, Ann Brobst, picked up her phone, dialed Bloodsworth, and rapidly set in motion the process that would finally put all any doubts to rest. She said she had some information to share with him and would only share it in person, at a place of his choosing in Maryland.

Bloodsworth asked to meet in the parking lot of a Burger King near his home. The next day, he showed up outside the restaurant up with his new wife, Brenda; a cousin; and his attorney. Brobst arrived in the company of a pair of police officers. The group went indoors. Brobst bought Bloodsworth a soda. Then she told him that investigators had finally linked the DNA to the real murderer, and she apologized.

There was one more surprise. The DNA belonged to a convicted sex offender named Kim Ruffner.

"I know him!" exclaimed Bloodsworth. Ruffner had been an inmate in the same prison where Bloodsworth had been held, in a cell just one level below Bloodsworth's.

◆ ◆ ◆

Bloodsworth, an honorably discharged ex-Marine, was the first person who had been on death row exonerated by DNA evidence in the United States. Ray Krone, an honorably discharged Air Force vet, had been a US Postal Service employee until he was tried and sent to death row. He would become the twelfth prisoner released thanks to DNA evidence (as well as the hundredth to be exonerated since the death penalty was reinstated in 1976 in the United States).

On December 29, 1991, the body of a thirty-six-year-old bartender named Kim Ancona was found in the men's room of the Phoenix, Arizona, bar where she worked. Evidence was minimal. Her body was disrobed, but there was no semen residue to be found. There were, however, irregular bite marks on her neck and breast, as well as traces of saliva. Krone played on a softball team the bar sponsored and visited the place regularly. Police investigators first took interest in him when one of the victim's friends told them that Ancona had mentioned that Krone might be helping her close down the bar that night.[7]

When they interviewed the then-thirty-five-year-old Krone, one of the police immediately noticed that his teeth were a mess: irregular, much like the impressions of jagged bites on the victim's body. Krone complied with a request to provide an impression of his teeth. He had an alibi. Krone owned his home and rented out a room to defray expenses. His male roommate told investigators that Krone had been home all evening. Krone claimed he was sound asleep well before the murder in the restroom took place. Nevertheless, based on what the Phoenix forensics lab decided was a match between the bite patterns and Ray Krone's irregular teeth, he was arrested and charged with sexual assault, kidnapping, and stabbing Ancona to death.

Ray Krone not only had no criminal record, but he'd been careful to never even get so much as a high school detention, since that would have interfered with playing on school sports teams.

The single piece of damning forensic evidence was the irregular tooth pattern. A key witness for the prosecution was Ray Rawson, a Nevada dentist who specialized in "forensic dentistry." Krone says he was "dumbfounded when this supposed expert, with all these credentials" testified that he was utterly certain that the molds taken of Krone's teeth were a perfect match for the bite pattern. Police apparently deemed it coincidental that hairs found and around on the victim were not Krone's.

Krone admits that he made a huge mistake in not finding a way to secure more competent legal counsel. He could have sold his house but didn't want to lose it. He says he "naively assumed" that his court appointed attorney, whom he later learned was paid only five thousand dollars, would be able to establish innocence.

It surely didn't help that by the time of the trial, Arizona news media had published lurid reports before and during the trial, dubbing Ray Krone "the snaggle-toothed killer." In 1992, he was convicted and sentenced to death.

It turned out that Krone had access to something many falsely convicted people don't: a caring relative with financial resources, a cousin in California who had founded a successful small software company and, after visiting Krone in prison, became firmly convinced Krone hadn't gotten a fair trial. In 1996, Krone's newly retained lawyers were able to get him a new trial after they had found potentially exculpatory evidence suggesting the bite pattern did not, in fact, match his teeth.

This time, the new defense had its own forensic experts. Three credentialed forensic dentists presented detailed evidence that that the bite mark on the victim simply was not a match for Krone's teeth, "snaggled" though both might be.

To Krone's astonishment, he was reconvicted anyway, based almost entirely on the testimony of the prosecution's original dental expert, an outright rejection of the testimony of the new defense expert witnesses. This time, though, the judge refused to sentence him to death, instead sentencing Krone to life in prison, citing his own doubts that Krone had committed the crime at all.

In 2001, a newly passed state law that opened the door for post-conviction DNA testing of allowed Krone's attorney to obtain a court order for testing of small samples of blood found on the victim's jeans and underwear. The same Phoenix police forensics unit that had helped get Krone arrested and convicted him this time determined that the DNA fingerprint in that blood was neither the victim's nor Ray Krone's, but someone else's. After checking an FBI DNA database, the Phoenix investigators did find a matching DNA fingerprint. The owner of that DNA was named Kenneth Phillips and had lived a few hundred yards from the bar where Ancona was killed. He had likely cut himself with the murder weapon during the struggle. Phillips was already in prison, convicted of choking and assaulting a seven-year-old child just a few weeks after Ancona's murder. Phillips, who at the time of the murder was out of prison on parole for another assault and choking, had never been considered a suspect even though he, too, had a terribly snaggled set of teeth. Confronted by investigators, the implicated killer at first denied the crime. But after more extensive questioning by a private investigator working with Krone's defense team, Phillips confessed to having woken up from a drunken stupor the morning after the murder, covered with blood, with no memory of the previous night. In 2006, he accepted a plea bargain and is now serving a life sentence for the murder.

Today, Krone's job centers directly on exonerating the innocent. He is a senior staff member of Witness to Innocence, a Philadelphia-based nonprofit organization focused solely on exonerating people falsely accused of crimes. Full members of the organization are an exclusive group: it's limited to people who have themselves been exonerated and are now free.

However, lest I mistakenly give the impression that forensic use of DNA is simple and straightforward, it's important to recognize how badly wrong matters can go if bungled or misapplied. Peter Gill, the forensic scientist who got his start in genetic forensics with the investigation the led to the conviction of the murderer Colin Pitchfork, is now a professor of forensics genetics at the University of Oslo, in Norway. In his 2014 book *Misleading DNA Evidence: Reasons for Miscarriages of Justice*, Gill points out that the original profiling techniques he helped Alec Jeffreys refine required relatively large samples, large enough to be visible to the naked eye, a minimum of about one centimeter in diameter. With the later development and refinement of polymerase chain reaction methods, scientists could suddenly work with increasingly microscopic samples, essentially blowing these "trace DNA" samples up into usable volumes by repeated replication.[8]

As Gill notes, at the trace level "DNA is everywhere in the environment." Consider this. Let's say you visit a gas station restroom a few hours after I do. It's winter, but I touched the door handle with bare hands. You're wearing gloves that touch the same entrance door handle I did. A few hours later, still wearing the gloves, you commit a robbery with a gun, discarding the weapon as you flee. A tiny bit of *my* DNA might well appear on the gun. And according to Gill, direct tactile contact isn't even required. I can transfer a tiny DNA aerosol not only by coughing in your presence, but even by merely *talking*.

One now-infamous debacle of an investigation—actually multiple investigations in multiple European nations over a period of years—shows how trace DNA can mislead. In 2007, a twenty-two-year-old policewoman was killed in Heilbronn, Germany. Police were not only able to locate trace DNA "evidence" at the scene, but they were also able to link that DNA to what appeared to be a serial murderer, one whose DNA evidence had been found at other crime scenes. This serial killer was a real rarity: a female. The DNA sample had two X chromosomes. Popular media quickly came to call her "The Phantom of Heilbronn." She was an odd criminal indeed: investigators were able to link her DNA to traces found all manner of crimes, including petty theft, burglaries, and car thefts in France and Austria as well as Germany. In a few of the more minor cases, apparent accomplices of various nationalities were prosecuted, but they all denied the female Phantom existed. After two years of furious searching for their Phantom, authorities discovered their woman. The DNA was indeed that of neither a serial killer nor a Phantom, but of an Austrian factory worker whose job it was to package the cotton swabs that police were using to collect DNA samples from crime scenes. The swabs were guaranteed to be sterile. But while sterilization kills disease organisms, it does not eliminate trace DNA.

As Gill emphasizes in his book, DNA fingerprinting can and did fail in one of the most tabloid-explosive crimes of modern times, the murder of British college student Meredith Kercher and the conviction of the young American college student Amanda Knox and her then-recent Italian boyfriend, Raffaele Sollecito, also a college student.

On November 1, 2007, Kercher was brutally murdered, killed with a knife, in the four-bedroom apartment in Perugia, Italy, that she shared with Amanda Knox and two young Italian women.

Several days after the murder, the DNA, footprint, and bloody handprint of young Côte d'Ivoirean drifter, drug dealer, and occasional burglar named Rudy Guede was found abundantly at the crime scene. Guede's DNA was found on Kercher's clothes, her purse, in the form of semen inside her, and in feces left unflushed in a bathroom. Guede had by then fled to Germany and was located after an extensive manhunt.

Guede was convicted and is in prison. But even before discovering the evidence that placed Guede in the room, police and, especially, "public minister" Giuliano Magnini had developed an elaborate theory about how Knox and Sollecito had committed the crime as part of a wild, drug-fueled satanic sex ritual.

The only forensic evidence the prosecution could point to implicating Knox came from one instance of trace DNA evidence. It's not clear exactly why police were initially suspicious of a large kitchen knife pulled from a drawer of knives, spoons, and other cooking utensils in Sollecito's kitchen. But since both Sollecito and Knox acknowledged that she had cooked several times with the knife, it was no surprise that some of her DNA was found on the handle. This came to be called the "double DNA knife" because police forensic investigators claimed that they had had also found an infinitesimal trace of human DNA on the blade that, after great amplification, they were able to match to Kercher.

However, it turned out that the knife could not have made the fatal wounds. The blade was too wide to match the size of the wounds at the appropriate depth. In addition, the bloody imprint of a much smaller knife appeared on a bed sheet at the crime scene. So the prosecution narrative was readjusted to suggest that the band of murderers had employed two different knives, the large one from the kitchen drawer, and a smaller one.

The sample that would be identified as Kercher's DNA on the knife blade was amplified using a controversial DNA fingerprinting technique called *low copy number DNA profiling*, in which a profile can theoretically be obtained by amplifying samples so tiny that they would be deemed inadequately small for routine analysis. In the United States, the FBI does not consider low copy number DNA profiling to be reliable enough to produce evidence admissible in a courtroom and does not allow any low copy analysis to be uploaded into its national DNA database.[9]

The prosecution's forensic investigator, Patrizia Stefanoni, had less than a hundred picograms (for a sense of how little that is, a paper clip weighs about one gram; a picogram is a gram divided into one thousand trillion equal parts) of material to work with. Only a few labs in the world are equipped to pull off the low copy number technique, which is so sensitive to contamination by infinitesimal bits of stray DNA that finds its way into the air or surfaces in the labs that labs using the technique must operate specialized equipment in cleanrooms with highly controlled ventilation. That didn't describe Stefanoni's rather basic set up. Nevertheless, when her machine kept reporting a sample size that was "too low," she overrode the machine's stated parameters to over-amplify the (supposed) DNA target. After she had finally succeeding in pushing her machines far beyond their designed limits of reliability, she finally got a result that led her to claim that she'd achieved enough amplification to identify Kercher's DNA on the knife.

No one double-checked the results. Nor will anyone ever be able to do so. The tiny trace was so thoroughly consumed in the procedure that reliable re-testing became impossible.

A number of qualified forensic DNA experts in the United States, the United Kingdom, Norway, and, notably, Italy have weighed in, sharply criticizing collection of evidence and its analysis in this case. First, only the one knife was taken from the drawer. A police officer said he picked it because it looked unusually clean to him, because it had scratches on the blade suggesting vigorous scrubbing, and because of "police intuition." Second, video documenting evidence-collection at the crime scene showed that police never changed rubber gloves, a standard protocol to prevent cross-contamination, meaning that DNA left innocently on, say, a door handle could easily be transferred to any other item the investigator later handles, with the police themselves serving as vectors of the DNA transfer from place to place at the crime scene.

Under a more standard and more rigorous, scientific protocol, other samples of would be retrieved to see if similarly tiny traces of "Kercher's DNA" appeared elsewhere as well, a sure sign that something might be amiss in collection or analysis: on another knife, say, or a can opener.

A request by the defense in the initial trial to have independent experts involved in the analysis was denied. The only review of the police

DNA expert's work was by her own boss who, like her, reported directly to the prosecutor. The boss said her work was "excellent."

A later appeals court did agree to a detailed review by outside DNA forensic experts, appointing two professors at the University of Rome: Carla Vecchiotti and Stefano Conti.

The short version: these experts tore the original police finding apart, issuing a damning report of the entire DNA analysis process, suggesting that there was no way to verify that what was tested was the victim's DNA and, even if it were, that there was no way to be certain it wasn't a tiny stray bit of contamination picked up during the investigation or in the lab, which would have been plausible because the same lab had been used to sequence Kercher's DNA.

A different single piece of DNA evidence also supposedly linked Raffaele to the crime: a clasp that appeared to have been cut from Kercher's bra during the attack. The clasp had remained at the crime scene for more than forty days before police finally thought to retrieve it as evidence. Photos show that it had been moved several times including, at one point, onto a pile of refuse. The university experts also tore into the analysis suggesting that some DNA found on the clasp belonged to Raffaele, noting, among other criticisms that "the international protocols for inspection, collection, and sampling of the item were not followed."[10]

Peter Gill catalogued some of those omissions:

> There was a 47 days [*sic*] interval between the discovery of the bra-clasp at the crime scene and its collection. During that time, the clasp had been moved and was found under a rug. In addition, there is video footage of the clasp being passed round police scientists, dropped on the floor—and the suggestion that gloves used to handle the evidence were not changed in between handling different objects. Shoe covers were not changed as investigating officers walked through the crime scene.[11]

The Italian justice system is dramatically different from that in the United States. Although there is presumption of innocence, the accused can be held for months without charges being filed, as was the case with

Knox and Sollecito. Although a couple of non-jurists sit on the jury-like panel, judges are directly involved in judgments and, in fact, direct the panel as it builds a conclusion based on evidence and "logic." Initial trials strongly favor prosecutors. On a more positive front, there is a liberal, virtually automatic appeals process.

The first trial, led to a guilty verdict. As part of a lengthy "reasoning" issued by the court, there was a statement that the *positioning* of Knox's DNA on the cooking knife's handle implicated her because it "appears more likely to have been derived from her having held the knife to strike, rather than from having used it to cut some food."[12]

For his part, Gill calls this "dangerous speculation" based on "confirmation bias," noting "there is nothing in the scientific literature that remotely supports such an inference."[13] Indeed, as he points out elsewhere, there is scientific evidence that trace DNA can move from one location to another on a piece of evidence, as in the case of a knife being transported: to the walls of the packaging in which it is being moved, as well as from blade to handle, and vice versa. In these experiments, knives were moved in cardboard tubes. In the case of the Kercher investigation, the suspect knife was moved in a shoebox of unknown origin.

In 2011, an appeal led not only to a not guilty finding but also to a second, stronger finding available under Italian law, outright innocence. The finding came with a scathing rebuke of the first trial's process. That finding focused heavily on what the appeals judge saw as the highly questionable DNA evidence. At that point, Knox and Sollecito were freed. She quickly returned to the United States.

There is, however, no double jeopardy rule in Italy. The two were retried and once again found guilty in early 2014.

But in the final episode of this story, in March 2015, the national Court of Cassation, what amounts to the supreme court of Italy, finally and definitively overturned that conviction on appeal, citing what it called the "stunning weakness" of the prosecution's case, including the "absolute lack of biological evidence" that either defendant had been in the room where Kercher was murdered.[14]

Steven Moore, a retired twenty-five-year FBI veteran began studying the case after seeing an early ABC *20/20* television news report that cast doubt on much of the "evidence" that had been reported in both tabloid and more mainstream media. In 2010, Moore said, in an interview on NBC's *Today*:

> In a crime scene like that, when you have so much blood, it's as if you threw blood all over the floor. If Amanda Knox and her boyfriend and that drifter were involved, there would be three sets of fingerprints, three sets of footprints, DNA, hair samples. It would have been a zoo of evidence. There was [*sic*], in that room, footprints, fingerprints, DNA, hair samples, saliva samples, everything for one person—a drifter. There is no way they [Knox and Sollecito] could have been in that room without their physical presence being obvious.[15]

There's no question that DNA-based analyses can be a powerful scientific tool in criminal investigations, helping both to identify the guilty and to exonerate the innocent. The Innocence Project, a New York–based nonprofit that relies on DNA evidence to help free those wrongly convicted of terrible crimes, points out that the effectiveness of DNA fingerprinting comes in great part because it was developed and refined by skilled scientists in peer-reviewed, high-quality research studies. That's in sharp contrast to forensics pattern-identification procedures such as the kind of bite-mark analysis used to convict Ray Krone, a technique that in 2009 the US National Academy of Science singled out as one of several forensic pattern-analysis approaches not backed by credible science.[16]

But just as with genetic tests in clinical medicine, for this tool to be effective, the science must always be rigorous and "not based," as Peter Gill puts it, "upon 'armchair arguments' about possibilities and theories."[17] Any gaps must be filled through careful experiments and the results fully disclosed. Gill notes that in the case of the Kercher murder, investigators had used a highly sensitive and precise test to detect any traces of blood on the kitchen knife, and had found none, a credible finding. They went on to make an unsupported claim that every last iota of blood DNA must have been scrubbed away with bleach, yet with one tiny bit of non-blood DNA from Kercher surviving that scrubbing.

Nothing in the scientific literature supports that idea, he notes. To credibly make such a claim, investigators would have to conduct carefully controlled experiments on a series of blood-covered knives to prove that such an outcome is possible, Gill writes, laying out a detailed seven-step process.

Whether the relevant parties are medical patients or judges, juries, and those accused of crimes, effective DNA analyses always demands scientific rigor and a determined focus on minimizing biases, along with full disclosure of any scientific limitations and uncertainties. For patients and their loved ones, anything less can lead to consequences ranging from needless anxiety to missed opportunities for treatment of profoundly serious diseases. In the criminal justice system, consequences can be the nightmare of unjust accusation, prosecution, and incarceration. In either arena, botched analyses ultimately can be a matter of life or death.

ELEVEN

The Decorated Genome

The brief but brutal famine that would come to be known as the Hunger Winter came to the western Netherlands in the waning months of World War II, the consequence of a late 1944 to early 1945 German army blockade of the region. Food supplies dwindled to a perilous third of normal, reducing the Dutch population in the region to scrounging for sustenance as mean as grasses and tulip bulbs.

So it came as no surprise that the babies of many of the mothers who had been pregnant during the Dutch Hunger Winter were born with lower than normal birth weights. The problem was especially acute with babies whose mothers were malnourished from the Hunger Winter during the late months of pregnancy, because that's when the developing fetus tends to pack on the most pounds.

In contrast, babies whose mothers suffered early in their pregnancy but then had adequate nutrition in subsequent months seemed to have "caught up," putting on enough pounds late during the pregnancy to be born at normal or near-normal weights.

Still no surprise to medical science. But major surprises from what amounted to a horrific experiment in maternal and fetal malnutrition would emerge in time, surprises that help cast light on how traumatic experiences like famine can alter our genes with changes lasting for many decades into our adult lives.

Thanks to the Dutch penchant for meticulous record keeping, as the 1940s rolled into the 1950s, 1960s, and beyond, epidemiologists were able to compare the children who survived the Dutch Hunger Winter to their unaffected siblings as they all grew into adulthood. As brothers and sisters, they shared half of their genetic code and turned out to be an incredibly useful study group to analyze the effects of famine on those deprived of adequate nutrition in the womb.

The good news was that the children who had been born during and after the Dutch Hunger Winter with low birth weights turned out to be fairly healthy adults. True, they remained shorter than average both as kids and as adults, but they also tended more often than their siblings to avoid becoming obese and were thus spared of all the attendant health side effects of obesity. Was this a fluke?

Not so for the other group of Dutch Hunger Winter children: the babies whose mothers experienced malnutrition in the earliest stages of fetal development, the very children who had "caught up" as fetuses when the food shortage subsided and thus were born as seemingly normal babies. Scientists who tracked their later development through childhood, into adulthood, then into middle age and beyond found an astonishing, disturbing, and statistically unlikely pattern. This group turned out to be inclined not only to obesity, but to the multiple-whammy of what is sometimes is called *metabolic syndrome*. They were not only significantly overweight, but also strongly inclined to diabetes, hypertension, and heart disease. Adults from this cohort had twice as much heart disease, for example, as groups of Dutch children from both preceding and subsequent years.

Clearly, this surprising outcome was not a matter of DNA, the province of genetics. The young victims of the Dutch Hunger Winter, after all, shared DNA with siblings born before and after them. They also shared with their siblings the "nurture" side of the nature/nurture equation, living in the same homes and eating at the same tables. Their siblings grew up without such a high rate of disease. What had happened?

Researchers eventually came to speculate that the answer to this puzzle actually *did* involve DNA, in a fashion. It wasn't that the master genetic code in this cohort had itself been altered. That much was clear. But somehow, the code had been decorated with a special ornament, a label that made the DNA read in a strikingly different way. By the year 2010, researchers at Holland's University of Leiden, working with colleagues at Columbia University in the United States, had worked out what looked like a potentially definitive cause.[1]

Of the twenty-thousand-odd genes, they had found particularly unusual findings with a gene called *insulin-like growth factor*, or IGF2, one already known to play key roles in early development and affect the weight and size of people even into adulthood. IGF2 is a hormone produced by the liver and deposited into the bloodstream, and is thought to be the major messenger for growth hormone and one of the most important proteins for determining body size and height. These scientists were looking in particular at chemical structures called *methyl groups* decorating those genes—compounds well known to play key roles in *epigenetics* (the first syllable from the Greek *epi*, meaning "above" or "beyond"), which refers to processes that do nothing to alter the genetic blueprint itself within each cell, but that can nevertheless dramatically alter how a given cell or group of cells biologically interpret that code.

The analysis paid off. They found that adults—now in their sixties—who had been in the womb at the most critical times during the Dutch Hunger Winter had many fewer of these methyl groups turning on and off the IGF2 gene than did their siblings who were born either earlier or later. This important work appears to have confirmed what had been speculated about for years: through the process of turning on and off genes like IGF2 in the Dutch Hunger Winter, babies had been altered epigenetically—again, a through a process operating beyond genetics itself—by a stressor, in this case malnutrition. Significantly, this mark on the DNA had persisted for a lifetime.

What's still uncertain is precisely why the clearly damaging effects of the Hunger Winter occurred. It could simply be that the epigenetic change somehow produced what University of Leiden scientist Eline Slagboom characterized as a kind of "scar" on the DNA. But an intriguing alternative is that evolution has somehow embedded this kind of response into the human *epigenome* as a response to famine: that it amounts to a sort programmed survival mechanism for a child likely to be born into a world of hunger and starvation. Another collaborator on the 2010 study, Bas Heijmans, suggested that the "epigenetics could be a mechanism which allows an individual to adapt rapidly to changed circumstances. . . . It could be the metabolism of the children of the Hunger Winter has been set at a more economical level, driven by epigenetic changes."[2]

The influence of parents' diet on their unborn children is sometimes known as Barker's theory, named for British scientist David J. P. Barker, who first proposed it in 1990. Barker's theory states that paternal diet, such as high fat or low protein, or maternal caloric restriction, prepare the fetus after birth for famine-like conditions, also called a *thrifty* phenotype, with a metabolism adjusted for optimal weight gain and retention. A corollary is that neonatal adiposity, or "baby fat," is higher in children of mothers undernourished during pregnancy. Similarly, Barker's theory proposes that human, as well as mouse, fathers who are undernourished tend to have children with higher rates of obesity. With mice, rigorous experiments have shown that restricting mouse fathers to low-protein diets will lead to obesity in their offspring, suggesting that the stress of the insufficient diet leads to epigenetic changes in their sperm cells.

Epigenetics itself is central to the normal development of animals and plants. After conception, as the first undifferentiated stem cells divide and produce new cells, a complete copy of the genetic code is packed into every successive daughter cell. How does the complete genetic code in one cell discern that it is to produce the proteins that will define it as a skin cell, in another a liver cell, another a brain or blood cell, or any of the hundreds of distinctly different cells with distinctly different functions? How, in other words, do new skin cells "know" to read only the pieces of the complete code that define skin cells, and to ignore the parts that have to do with brain or blood cells?

The answer is by the action of methyl groups and a few other types of epigenetic chemical alterations that act something like switches, clicking on or off pieces of the full genetic blueprint in each cell, a process that's been compared to highlighting selected words and sentences in a vast encyclopedia, and blacking out others.

The DNA code, in other words, isn't altered at all. But these "beyond genetics" switches can alter the way the code is read. In each cell, the long, double helix of DNA efficiently packages itself by winding in complex loops around structures called *histones*, which function somewhat like spools for thread. How tightly or loosely the strand wraps itself around a histone in any given cell helps determine which bits of code are turned on (expressed) and which are turned off (suppressed). Additionally, a process called *methylation*, in which patterns of methyl or larger acetyl groups that attach to the DNA, further controls gene expression.

Considering how critical epigenetics are to normal cell development and differentiation, it's no surprise that its abnormalities can lead to dysfunction and disease. In fact, scientists are trying to learn more about the role epigenetics play in a host of diseases, from autoimmune disorders like lupus to cancer. One intriguing line of research suggests that genes that might normally suppress cancer are shut down in some tumors by *methylation*. This appears to be detectable early in the development of some kinds of tumors, including colon cancer, and could be key for developing better tests for the disease.

Surprisingly, emerging discoveries suggest that epigenetic changes can actually be inherited across generations, upending the conventional notion that inheritance can only happen via transmission of DNA from parents to their offspring.

Indeed, epidemiologists following the Dutch Hunger Winter cohort report they have "preliminary evidence" of the pattern of poor health extended all the way to the grandchildren of the women who had first suffered famine, children never themselves exposed to that nutritional stress.

Studying these kinds of effects in humans is daunting. Humans have long lives and are far more genetically diverse to start with than laboratory fruit flies or mice. We also (for peculiar reasons that researchers don't understand) have a tendency to not want to participate in experiments for decades that restrict what we eat, where we go, and particularly who we choose to mate with.

But scientists working with laboratory mammals have also seen intergenerational effects. Consider the case of the fruit fly *Drosophila melanogaster*. In 2009, scientists at the Swiss science and engineering university known, in the English speaking world, as ETH Zurich (in German, Eidgenössische Technische Hochschule Zürich) reported that they had been able to alter eye color in the tiny flies by exposing embryos to unusually high temperatures. When the researchers, led by biologist Renato Paro, raised temperatures from a normal 25 degrees Celsius to 37 degrees Celsius, the offspring were born with red eyes rather than normal white eyes. Paro and his team then crossed and re-crossed successive generations of red-eyed flies, exposing these offspring to perfectly normal embryonic temperatures, and found that the effect persisted through at least six generations. When they sequenced the DNA of the red-eyed offspring, they found it to be an unaltered code: the same DNA as the original parents.[3]

Consider, also, the case of the gene dubbed *agouti* in mice. If a mouse that should have fur a normal brown color is born without a specific pattern of methyl groups on this gene, it will have yellowish fur. These mice also typically become obese and are more prone to cancer. Various experiment have found that if a pregnant female with this disordered gene is fed a diet rich in folic acid, vitamin B12, and the nutrient choline, (which are the building blocks for DNA methyl groups in cells and so likely enhance this epigenetic change) most of her offspring will be normal brown pups and will develop normally, that is, not become obese. When normal-looking females mature and themselves are bred, their pups tend to be normal and have higher amounts of methylated DNA.[4]

A 2013 article in the journal *Biological Psychiatry* focused on stress that appears to be transmitted inter-generationally. Researchers at Israel's University of Haifa had previously shown that exposing female lab rats to even "minor" stress before they conceived could be linked to behavioral changes in offspring. The minor stress amounted to changes in temperature of the rats' environment and in their daily routines, compared to a control group that lived in a constant temperature with a regular routine. The stress came when the rodents were forty-five days old, essentially in the rat developmental age of adolescence.

Working with professor Micah Lesham, who had conducted the original stress studies, PhD student Hiba Zaidan repeated the experiment, focusing on a gene called CRF-1, which was already known to express itself in the brain in response to stress. Before the animals were bred, the researchers found that the gene was more active in the ova of the previously stressed females than in the non-stressed mothers. Consistent with this, they also checked the DNA in brain tissue of females' pups immediately after they were born and found greater expression of CRF-1 there among the stressed group as well.

In a third step, the scientists exposed some of each group of the offspring, once they had matured, to the same types of minor stress applied in the original experiments. The result: female offspring (but, curiously, not males) whose mothers had been stressed showed markedly higher CRF-1 expression. "If until now we saw evidence only of behavioral effects," they noted, "now we've found proof of effects at the genetic level," at least in rodents.

Pointing to the "systemic similarity" of rodents to humans, the scientists also suggested that the study could well be relevant to human mothers exposed to various stressors and their children.[5]

Even earlier, in 2012, scientists Michael Skinner, at Washington State University, and David Crews, at the University of Texas at Austin, published a study in the *Proceedings of the National Academy of Sciences* suggesting that exposing pregnant rats to a commonly used fungicide called vinclozolin appeared to increase stress reactions not only in a second generation of offspring, but into a third generation.

Although Skinner had been studying epigenetic effects of pollutants, pesticides, and other chemicals for years, he says, "We did not know a stress response could be reprogrammed by your ancestors' environmental exposures."

Crews adds, "We are now in the third human generation since the start of the chemical revolution, since humans have been exposed to these kinds of toxins," suggesting that "this is the animal model of that."[6]

It's well known that the food or chemicals a mother is exposed to can affect her children's health. But it is a paradigm shift to suggest that the environmental or even psychological milieu either parent experiences can affect, profoundly in some cases, not just one's children but successive generations. If true, it would amount to a major change in how we need to think about how our experiences affect not only us, but can have effects reaching far into future generations.

TWELVE

The Age of Geneticism

Aldred Scott Warthin was a brilliant and driven man. A lifelong lover of learning, in 1891 and 1893 he was awarded his MD and PhD degrees, respectively, an almost unheard of academic accomplishment at the time that put him among a rarefied group of the world's most elite intellectuals and scholars. Eager to make his professional mark on a vigorous, rapidly maturing, and ambitious academic medical culture in the United States that was seeking to compete with its European counterpart, in 1894 Warthin became director of the pathology laboratory at the University of Michigan School of Medicine.

His meticulous examination and evaluation of patients gave the world what is called *Warthin's Sign*. That's a particular scratchy tone doctors hear when listening to the sounds of their patients' hearts and lungs through a stethoscope.[1] It is an indication of pericarditis, an inflammation of the tissue surrounding and protecting the heart. He was also the developer, with a colleague, of the Warthin-Starry stain, a chemical test still used today for the presence of the particular type of bacteria called spirochetes that cause syphilis.[2] Warthin authored more than a thousand scientific papers and several landmark textbooks in pathology. His scientific work ranged from studies of venereal disease to the role of lymph nodes in the immune system. During World War I, he served as a key consultant to the US government on the effects of poisonous mustard gas, one of the earliest chemical weapons (and used by both Allies and Central Powers). He and Sir William Osler of Johns Hopkins University were widely considered to be the two giants of medicine in the late nineteenth and early twentieth century.

Additionally, Warthin was a polymath. He had studied science at the University of Indiana before moving on to medical and graduate school at Michigan. But even before that, he had been in love with music. Warthin had graduated from musical conservatory in Cincinnati with degrees in both piano and violin, and went on to tour Europe, playing music at Freiburg and Vienna. On a trip to California during his later academic medical career, he identified a new species of snail that was later named after him. He was also an avid gardener who judged botanical shows in Detroit. Along with his volumes of scientific publishing, Warthin wrote several books on philosophy, art, and music.

In 1926, he wrote about the importance of physicians getting daily exercise: "But if we have not developed a varicose type of physician through the increased sedentary character of his daily routine, we certainly to some extent are developing an automobile type of overprosperous appearing, obese, albeit flabby, somewhat short-in-the-wind, slightly cyanotic and prematurely bald human being."[3]

As a teacher in the medical school, Warthin was also very demanding of his students and not the most tolerant man. As an assignment to help his medical students identify heart murmurs, by teaching them how to listen carefully and focus on detecting subtle acoustic changes, he had his medical students memorize the tones of Ann Arbor church bells and the university's carillon. He once blamed the spread of syphilis on the doctrine of religious forgiveness. He was an avowed eugenicist, urging people to reject traditional religion and ground their lives with faith in modern science.

"Man needs a new religion if the race is to be saved from degeneration," Warthin wrote. Favoring eugenic marriage, Warthin startled the country with the comment that "love was emotional nonsense," at the third Race Betterment Conference, in Battle Creek, Michigan, in 1928. He told young men to investigate the medical histories of their prospective spouses "and not to hesitate to reject them if freedom from inheritable diseases could not be proven. . . . The human race could never win its way to physical betterment or longer life until all marriages had a eugenic basis," he wrote.[4]

When Warthin and his wife first arrived in Ann Arbor (they had two daughters and two sons), the couple settled down in a traditional stone house on Ferdon Street, not far from the University of Michigan Hospital. As was typical in a time before dishwashers and washing machines, Warthin hired several people to help his wife run his household: a cook, a maid, and so on. Pauline Gross was a seamstress hired by the Warthin family to help mend the family's clothes. She was reported to be a pleasant but somewhat serious-minded seamstress. Warthin needed his clothes to look professional and proper during his lectures, on medical rounds, playing golf and his many eclectic other pursuits. It turned out that Warthin was so active and vigorous that Pauline Gross was having trouble keeping up with her mending work for the household. This displeased Warthin greatly.

One day after a long day at the Hospital, Warthin confronted the woman about her work for his family. The seamstress admitted that constant worry was interfering with her work.

"I'm healthy now," the seamstress said, "but I fully expect to die an early death from cancer. Most of my relatives are sick, and many in my family have already passed on."

Warthin a trained pathologist who had dissected the cadavers of many who had died from various abdominal cancers, found her comments thought provoking. That cancers or other inherited diseases could run in families was a novel concept to medicine at the time. But instead of dismissing Gross's concerns, this deeply intellectually curious physician wanted to know more.

Was this her depression talking, or could this startling tale be true? How could cancer be inherited from your family? As he questioned Gross further, she revealed that family members had come down with several different kinds of cancers—ranging from stomach, to rectal, to "female organs," and others[5]—which look very different under the microscope, and in many ways act very distinctly. This seemed very odd indeed.

Consider that Warthin's conversation with Gross came only thirty years after Gregor Mendel performed his classic series of experiments with pea plants that first outlined the laws of genetic inheritance, and before the proof of the existence of genes in simple animals like fruit flies. With regard to the practice of medicine, it would be another decade before the physician and scientist Archibald Garrod would describe the first well-characterized human genetic disease: *alkaptonuria*, a metabolic disorder that involves an inability to break down the amino acids phenylalanine and tyrosine.

Pauline Gross told Warthin that she and her family had lived in Washtenaw County since before the Civil War. Warthin took the names of her family members who had died and searched for their death records. He also contacted her living family members to corroborate and cross-check different versions of their stories, which is basically what a scientist does in the lab when he discovers a potentially significant but unanticipated and non-intuitive new finding. He spent many long days and nights working on this new project.

Warthin learned that Pauline Gross's grandfather was born in Plattenhardt, Germany, in 1796 and immigrated to the United States in 1831. Pioneer G, as Warthin called him, cleared woodlands and cultivated crops and started a family. By age sixty, he died from cancer of the intestines. He had five sons and five daughters. Four out of his five sons also died from cancer, as did two of his five daughters. His grandchildren numbered seventy, of whom thirty-three died of cancer, many at a young age. His children planted their family tree across two continents, some remaining in their native Germany, while others settled in the American Midwest in search of new opportunities and land. Some of the children of the Midwestern contingent had also migrated to Northern California during the Gold Rush to make their fortunes.

The family developed not just any cancers, in a random manner, but specific types: cancers of the colon, stomach, uterus, and ovary. This was the first description of what later become known as the cancer family syndrome. But two branches of the family remained essentially cancer free: the descendants of two cancer-free daughters.

Both intrigued and confused by his finding, Warthin devoted time to search for other families with histories of cancer at the University of Michigan hospital, as the medical literature had very little on this topic to report. He and his colleagues analyzed the medical charts of surgery patients from 1907 to 1909 for cancer diagnoses at the hospital. This study, however, showed that less than 1 percent of the cancer surgery patients there had described, as recorded in their medical histories, other family members suffering from malignancies. A determined Warthin nonetheless pursued this line of inquiry further. He wrote letters to patients and their families and even made house calls to hear their stories in person. These investigations showed that cancers in fact occurred in the kindreds of more than 50 percent of the hospital's cancer surgery patients. Warthin's interviews led him to conclude that this underreporting in the medical record was not accidental.

"Many have a certain horror or a fear of stigma attaching to a family history of multiple incidence of neoplasm," Warthin wrote in his notes on the family. Warthin originally thought that the cancer in Family G was recessive because about one-quarter (23.6 percent) of family members across the generations developed cancer by age twenty-five, although we now know that it is a dominant disease but that not everyone who inherits a mutation develops cancer.

Tragically, Pauline Gross herself later died of cancer, as she had both prophesized and feared. At that time, there was no medical screening intervention she could use to reduce her risk of dying. Warthin's work with the Gross family spanned almost twenty years. During that period, he used rigorous scholarship and hard facts to articulate what was at that time a thought-provoking and innovative concept: that cancer could be heritable in humans. Warthin's scholarly study was eventually published, in 1913, in the journal *Archives of Internal Medicine* as "Heredity with Reference to Carcinoma: As Shown by the Study of the Cases Examined in the Pathological Laboratory of the University of Michigan, 1895–1913."[6]

However, decades would pass before this landmark study would gain the recognition it deserved. Perhaps that's because the idea of heritable cancers was, in its time, so far out of the mainstream of medical thinking. Nor did very many people pay much attention to the work of Carl Weller, Warthin's intellectual successor at the University of Michigan, who would continue Warthin's work.

Likely inspired, in part, by his studies of what he came to call Family G, Warthin began to promote eugenics as a public health measure to reduce the cancer burden.[7]

Decades later in the 1960s, when a different mindset was permeating American society, the young and energetic Henry Lynch (described in chapter 7, "The DIY Genome") picked up the thread that Warthin and Weller had first spun. Lynch lived and worked in Omaha, Nebraska.

Sometimes treating several generations of the same family from their traditional ranches and farms, Lynch would see several members come down with the same constellations of malignancies: breast and ovary, colon and uterus, melanoma and pancreas. He had to dig deep into the stacks of medical literature, but eventually he found and read the original 1913 Warthin work in the *Annals of Internal Medicine* and noted striking similarities between Family G and the patients in their beds right in front of him. He lectured to the American Academy of Physicians, in 1965, and even met Warthin's grandson, who was a physician in Boston.

Like Warthin, Lynch would turn out to be ahead of his time. By the middle of the twentieth century, science had fully embraced the idea that environmental exposure to pollutants, toxic chemicals, radiation, and the like caused cancer. In fact, environmental exposures (including radiation from the sun, in the case of skins cancers) *are* often key triggers to the development of various kinds of tumors. What scientists failed to appreciate as recently as the 1960s was that cancer is a disease of the DNA, and that the very reason some environmental exposures cause cancers is because they initiate what becomes runaway damage to DNA. They also failed to appreciate that mutations in DNA inherited across generations could also be a cause of cancer.

"No one would believe me. I tried and tried and tried but couldn't get a grant to study how cancers run in families," Lynch recalled.

Lynch tells the story of one National Institutes of Health (NIH) grant he wrote on families that had hereditary breast-ovarian cancer (HBOC) syndrome. This is the syndrome now known to be caused by mutations in the now infamous BRCA1, BRCA2, and other genes. At that time, most NIH grants required in-person site visits by several reviewers (with subsequent positive opinions rendered) before a study could be funded by the federal government.

Epidemiology is the study of how environmental factors influence disease risk: for example, how smoking or asbestos exposure can affect a person's chances of developing lung cancer. For this study review panel, the NIH sent three prominent academic epidemiologists to Lynch's Omaha lab.

Lynch had put together a slide presentation to go over the project's background preliminary data. "While I was giving this presentation," he says, "I noticed one of the visiting epidemiologist reviewers kept looking at his watch impatiently. Then, in the middle of my slide presentation, he stood up while I was still talking and went in to ask my administrative assistant to call the airline and change his plane ticket, so he could get on an earlier flight out of Omaha. He told her, and later me, he had heard enough already to know that I clearly had no training in the field of epidemiology!"

During a site visit for another grant, Lynch presented his group's studies on the families that turned out to define what later was called the FAMM syndrome (familial malignant melanoma and pancreas, subsequently shown to be caused by mutations in the CDKN2A gene).

"At least these NIH reviewers were polite enough to stay until the end of our presentation," he says. After the group had lunch and finished the grant review site visit, Lynch drove the reviewers to the airport himself.

He reports that, before getting out of the car, one of them, a famous cancer epidemiologist, looked at Lynch, shook his head, and sighed apologetically. "You know," the epidemiologist said, "these relatives with aggregations of different cancers are all just coincidences. Whatever you say, I just don't believe cancers run in families."

He then suggested Lynch think about rewriting the study focus on environmental pollutant exposures to which these patients might have been exposed or foods they might have eaten.

In his book *The Structure of Scientific Revolutions*, science historian Thomas Kuhn described how both scientists and society come to see the world through the prism of prevailing paradigms. A classic example of how difficult it can be to shift out of faulty paradigms came when Copernicus and Galileo reported their findings that the Earth rotates around the sun, when the prevailing certainty was that the sun (and the planets and the stars) rotated around the Earth. After what came to be known as the Galilean revolution, the stars moved across the night sky in exactly the same way as before. The data recording their celestial movements across the nighttime sky scratched into papyrus were just as accurate as they had been before. But with the paradigm shifted, and the Earth and human beings no longer sitting at the universe's center, the existing observations about the movements of the heavenly bodies could be reinterpreted in the context of an accurate version of how the cosmos works.

Now, in biology and medicine, we are similarly in the midst of a genomic revolution that is shifting our paradigms of disease and the way we think about its root causes. Using the retrospectoscope as our prism, it is easy to look back at the times of Warthin and Lynch and think how silly it seemed for their contemporaries to participate in a collective *folie* and to not have immediately accepted the paradigm that genetic mechanisms learned from fruit flies and other animals could explain the etiology of some diseases running in families.

Yet today, people (including scientists, doctors and, dare I say, even genetics professionals) are not so different from generations past, or even from the contemporaries of Ptolemy two thousand some odd years ago. Even with our great technological sophistication and greater scientific knowledge of the universe, we are psychologically just as susceptible to *thought herding*. Thought herding is a common behavior where our assumptions about the world are all too quickly reinforced by those around us, who accept the same paradigms as rock-solid truths, immune to challenge or skepticism, and we in turn strengthen the faith of those surrounding us.

Today, even as someone who is immersed in genetics and loves it, I've seen enough evidence that perhaps the pendulum may have swung back too far with regard to our faith in genetics and sequencing genomes as the answer to all our personal medical problems.

I recall one patient, Jules, who was a physician's assistant. Jules had a bona fide, rare genetic syndrome that can trigger a type of *porphyria*, a cluster of diseases that mainly involve a buildup of excess compounds called *porphyrins*. Porphyrins themselves are very necessary precursor chemicals involved in production of hemoglobin, the molecules that ferry oxygen from the lungs to the body's cells. But when a large excess of porphyrins triggers porphyria, patients can suffer symptoms that range from aches and pains to nervous system disruption, even acute paralysis. However, people with these genetic syndromes sometimes go through life symptom-free as well. Porphyrias occur in about one in eight thousand people.

Jules told me why he had made an appointment with me. He needed to have his genome sequenced because he thought it likely that he had an *additional* rare genetic disorder, one that caused café au lait spots (which are dark, irregularly shaped birthmarks) on his body. Jules told me he had searched for his symptoms online. He was here to see me because the web search had led him to believe that he had the genetic disease *neurofibromatosis type 1* (caused by mutations in a gene called *Neurofibromin*), a disease that occurs in only about one in every three to four thousand people.

Jules wasn't finished. He next told me that even though it wasn't in his medical record, he also carried a third genetic disease, *hemochromatosis*, one of the more common recessive diseases, present in about one in two hundred people. It causes the body to absorb too much iron from the diet.

This genetic trifecta, he'd concluded, amounted to a new metabolic syndrome that explained his attacks of abdominal pain every few months, sending him to the emergency room for acute therapy and hospitalization.[8]

The story to me seemed incredible! Three rare Mendelian genetic diseases all diagnosed in the same patient? This was unheard of, like getting hit by lightning three times in the same day. A quick Medline search of the medical literature came up with, well, absolutely no precedent for this tri-genic constellation before. The thought went through my head that this person could have some sort of never-before-described mutator syndrome that caused multiple genetic diseases in the same person. Had I just encounter a brand new genetic syndrome? This was really exciting. Jules would be the first Lipkin syndrome patient! Anyway, that's how I imagined—for a few seconds—this remarkable new genetic mutator syndrome would be named.

However, I began to get a bit suspicious when I asked Jules if I could get the information from the laboratory testing he'd had at another hospital's ER during those painful attacks so that I could gain a more granular understanding of what diseases we were dealing with. Initially enthusiastic, he was now hesitant about releasing his medical records from other hospitals. Since Jules had been so very forthcoming about his medical history and presumptive diagnoses, this seemed odd.

With my skepticism growing, during a physical examination I studied carefully his dark, irregularly contoured birthmarks, and then asked gravely, "Do the centers of your café au lait spots ever turn spotted blue and the surrounding part of the birthmark turn pink?"

"Yes," he said, eyes widening. "They do! That's exactly right!"

Well, no, they don't. I had laid bit of a trap. The centers of café au lait spots do not turn spotted blue while the outer part turns pink, in any genetic or nongenetic disease. While Jules truly did have porphyria, he was not telling the truth about the other conditions.

An old saying in medicine goes "Common things occur commonly." That is, your doctor is more likely to find some unusual variant of a known disease, rather than a new disease entirely. And I suddenly had a pretty clear idea of what was really ailing Jules. It seemed likely that in addition to his one known rare genetic disorder, he had a second condition far more commonly reported by physicians and psychologists: Munchausen syndrome.

This is a mental disorder named for the Baron von Munchausen, an eighteenth-century German known for his extravagant exaggerations of his exploits as an army officer. Patients with the condition (which as of yet does not have a known genetic trigger) behave as if they are seriously ill, when in fact they are not. In part, this is thought to result from a desire to get attention.

I believe Jules may have been impressed with his first genetic diagnosis and sought more genetic testing based on his sense that learning more about his genes would be a solution to all of his medical problems, whether they were indeed genetic or not. Needless to say, I did not want to sequence this man's genome without good reason, for fear of what further psychological needs for attention it might evoke.

I recall another patient—let's call her Angela—who is an athletic, very intelligent, and accomplished executive in her thirties without any active health issues or notable personal medical history. However, when I took her family history, she told me she had an uncle who suffered a stroke in his forties. Angela's uncle had gotten his exome sequenced as part of a research study by a laboratory that was very enthusiastic about the power of genetics to identify previously undiagnosed illnesses, but perhaps less experienced at interpreting genetic variants as benign polymorphisms or disease-causing mutations. While exome sequencing did not reveal the cause of her uncle's stroke, it did reveal two unrelated variants of uncertain significance in genes that had been associated before with childhood heart congenital anomalies (unrelated to stroke risk), even though no one in her family had previously had these diseases. Additionally, four variants of uncertain significance in well-established cancer risk genes (three associated with hereditary breast cancer and one with familial colorectal cancer) were also discovered in her uncle, even though no one in their kindred had developed breast or colorectal cancer. A healthy, hardworking woman deeply engaged in her career and family, Angela came to me now upset and distracted as someone who suddenly was at risk of having heart disease, developing multiple types of cancer and passing these on to her children. After I examined her, her records and ordered further genetic testing and analysis by another, more experienced molecular diagnostic laboratory, it became clear that she was not at risk for any of these diseases. But the potential was there for overdiagnosis that threatened to transform a once happy, healthy woman into someone who saw herself as sick and was unnecessarily anxious about her future.

An oncologist colleague at the University of Michigan hospital once told me of another case of genetics-centered overreach that occurred there. A young woman in her twenties who was developmentally delayed, deaf, and institutionalized was transferred to the hospital from her nursing home for an unusual presentation of seizures. She had been restrained on her hospital gurney, out of fear that she might injure herself. Unable to speak or understand her doctors' questions, the disheveled patient slowly writhed. In an irregular cadence, she would shift back and forth across the bed, her head turning side to side, as her legs moved up and down on the sheets as if she was trying to walk while lying flat. Her lab tests and imaging could not explain her puzzling presentation. The attending physicians reasoned that she must be having an atypical seizure, so a neurologist was consulted. He examined her and found nothing particularly remarkable. The electroencephalogram he ordered showed no seizure activity.

This patient was clearly a black box with no key. One of the puzzled physicians guessed that she might have some rare Mendelian genetic syndrome, so the hospital's clinical genetics unit was called to see if they could make the diagnosis. A well-known and respected clinical geneticist (who wished to remain anonymous in order to avoid publicity and focus on his work) read the patient's chart and talked with the other doctors. When he arrived at her bedside, he sat down and, for a time, simply observed the writhing patient. Then he stood, walked to the nurses' station, and asked for a pair of tweezers and a cup. One nurse, a bit surprised, went to look for tweezers in the suture kits. She returned and gave them back to the geneticist, who went back into the patient's room. Everyone outside the room was puzzled. A few minutes later, he came out with the cup in his hand and sat at a microscope. Under the microscope he saw, on the slide he had quickly prepared, was the mite *Sarcoptes scabiei*, commonly referred to simply as scabies. These parasites are spread from person to person, and through dirty bedding and clothing. Inhabitants of medical institutions are known to be at increased risk of being infected with these mites. They can burrow under the host's skin and cause intense allergic itching, often at the wrists and ankles.

Common things occur commonly, and this was no unicorn among horses. The patient did not have a rare genetic syndrome causing developmental delay, deafness, and seizures. She was institutionalized and simply had scabies. She may have had some sort of genetic disease causing her developmental delay, but that was beside the point. Genetic causality has to be viewed in the context of the surrounding environmental circumstances and wasn't the real reason the patient had to be sent to the emergency room.

In the early twentieth century and again in the 1960s, we had many trained physicians and scientists psychologically unwilling to break away from the contemporary mainstream thinking and see the evidence (clear in retrospect) produced by pioneers like Warthin and Lynch supporting the importance of genetics in medicine.

Fast forward. Now in recent years we've seen full-scale media blitzes spotlighting single genes that are supposed to explain individual destiny in not-so-clearly genetic contexts ranging from homosexuality, to left-handedness, and even violent criminal behavior. The onslaught of news media reports seem to have conditioned many of the intelligent, accomplished individuals who come to me as patients as having already solved in their minds without any doubts the equation: genes = destiny without a healthy dose of skepticism. For some, genetics has become a default and sometimes fatalistic explanation for their medical, and indeed, life problems, which can minimize the importance of personal lifestyle choices to influence their own destiny and allow for rationalization of choices later regretted.

I have on more than one occasion had patients come to me and basically implore me, "Please sequence my genome," before they can tell me what is bothering them in the first place. Perhaps because contemporary Americans are often also believers in the doctrine that technology can solve all our problems, it can take a lot of time and pointed discourse to convince them that genome profiling may not explain all we would like to know about their medical destiny and future of their bloodline. In these instances, I wonder if faith in genetics has, for some, replaced faith in astrology as a tool to predict their future. Clearly, we must take a more balanced view that acknowledges the power of genetics to personalize our medical care, but avoids the reductionist misconception that applying genetics to medicine is a panacea that can precisely predict the future or solve all our family's problems with a diagnosis and a pill. This issue may be particularly acute because of the potential for genetic overdiagnosis to drive up medical costs, which could cause a deficit in social justice for some very expensive genetic medicine therapies.

A final reminder for patients: genetic medicine's goals are to make more precise evaluations of disease risk, a concept embedded in probability, not certainty. Unless you are dealing with one of the extremely rare cases of genetic diseases that are fully penetrant (that is, having a 99 percent chance of manifesting if you carry a mutation, such as for Huntington disease), if a doctor or genetic professional tells you he or she *knows* what is going to happen, find another health-care provider!